Methane Release—from Old Subsurface Origins to Recent Biogenic Production

Methane Release—from Old Subsurface Origins to Recent Biogenic Production

Guest Editors

Gabriel Barberes
Nuno Lamas Pimentel

MDPI

Basel • Beijing • Wuhan • Barcelona • Belgrade • Novi Sad • Cluj • Manchester

Guest Editors
Gabriel Barberes
Center of Geoscience
University of Coimbra
Coimbra
Portugal

Nuno Lamas Pimentel
Instituto Dom Luiz (IDL)
Universidade de Lisboa
Lisbon
Portugal

Editorial Office
MDPI AG
Grosspeteranlage 5
4052 Basel, Switzerland

This is a reprint of the Special Issue, published open access by the journal *Methane* (ISSN 2674-0389), freely accessible at: https://www.mdpi.com/journal/methane/special_issues/nature_methane.

For citation purposes, cite each article independently as indicated on the article page online and as indicated below:

Lastname, A.A.; Lastname, B.B. Article Title. *Journal Name* **Year**, *Volume Number*, Page Range.

ISBN 978-3-7258-6876-6 (Hbk)
ISBN 978-3-7258-6877-3 (PDF)
https://doi.org/10.3390/books978-3-7258-6877-3

Contents

About the Editors

Gabriel Barberes

Gabriel de Alemar Barberes is a Geoscientist whose expertise is rooted in a deep understanding of geochemistry and petroleum systems, honed through his Ph.D. from the University of Coimbra and significant R&D experience. He maintains a dynamic presence in both industrial and academic research, holding concurrent positions as a Researcher at Federal Fluminense University (Brazil), at the Centre for Geosciences (Portugal), and Geologist/Geochemist at the Petrobras Research Center (CENPES) in Brazil. His work focuses on applying advanced analytical techniques to complex geological challenges, from surface geochemistry for hydrocarbon exploration to the characterization of organic matter. A dedicated contributor to the scientific literature, he has authored several peer-reviewed articles, developed a patent for Raman spectroscopy data analysis, and guided scientific discourse as an Editor for journals such as *Geosciences*, *Methane* and the *Journal of Carbon Research*. This editorial experience is built upon a strong foundation of community service, evidenced by over dozens of peer reviews performed for international journals.

Nuno Lamas Pimentel

Nuno Lamas Pimentel has been an Assistant Professor at the Earth Sciences and Energy Department of the Lisbon University (Portugal) since 1998. With a PhD in Sedimentary Geology and a full-time academic, he has been working in the oil and gas industry since 2005. His industry-based research projects have focused on conventional and unconventional petroleum systems in Portugal and Brazil. Besides publishing scientific research, he has also been involved in the promotion of field courses for Petrobras (Brazil), GALP (Portugal), and REPSOL (Spain). He has authored, co-authored, and reviewed a large number of technical and scientific papers, the majority of which are highly ranked international papers, related to different oil and gas basins in Portugal, Brazil, Australia, Iran, Pakistan, and Bangladesh.

Preface

Trace gases constitute a minor fraction of the Earth's atmosphere, yet they exert a disproportionate influence on the global climate system. Among these components, methane stands out as a greenhouse gas of exceptional potency, presenting a complex challenge for the scientific community. Its presence in the atmosphere is the result of a dynamic interplay between ancient geological processes and active biological cycles. Understanding the full spectrum of these emissions is not merely an academic exercise but a prerequisite for accurate climate modeling and effective environmental stewardship.

The motivation for compiling this Reprint lies in the urgent need to bridge the gap between detection and mitigation. Methane occurrences are diverse, ranging from gas hydrates trapped in the subsurface to the enteric fermentation processes occurring in livestock. Each source presents unique difficulties regarding quantification and management. While geological seepage requires advanced imaging and monitoring technologies, biogenic production demands a deep understanding of genetics, microbiology, and animal nutrition.

This Reprint aims to provide a comprehensive overview of these distinct yet interconnected fields. It consolidates research that addresses the mechanisms of methane release, the technological innovations required to measure it—from micro-scale tomography to field-based laser detection—and the practical strategies available to mitigate its impact. By bringing together studies focused on subsurface origins and recent biogenic production, this Reprint highlights the multidisciplinary effort required to refine global methane budgets.

This Reprint is addressed to researchers, environmental scientists, animal nutritionists, and policymakers who are navigating the complexities of greenhouse gas emissions. We hope that the methodologies and findings presented herein will serve as a valuable resource for developing sustainable solutions to limit the rise in global temperature.

We extend our gratitude to the authors who contributed their expertise to this volume and to the reviewers whose diligence ensured the scientific rigor of the published work.

Gabriel Barberes and Nuno Lamas Pimentel
Guest Editors

methane

Article

The Development of a Low-Cost Method for Monitoring Methane Leakage from the Subsurface of Natural Gas Fields

Muhammad Alfiza Farhan [1,2,*], Yuichi Sugai [2], Nuhindro Priagung Widodo [1] and Syafrizal Syafrizal [1]

1 Department of Mining Engineering, Bandung Institute of Technology, Bandung 40116, Indonesia; agung@mining.itb.ac.id (N.P.W.); syafrizal@mining.itb.ac.id (S.S.)

2 Department of Earth Resources Engineering, Kyushu University, Fukuoka 819-0395, Japan; sugai@mine.kyushu-u.ac.jp

* Correspondence: alfizafarhan@rocketmail.com

Abstract: The leakage of methane from the subsurface on the coalfield or natural gas field invariably becomes an important issue nowadays. In notable addition, materials such as activated carbon, zeolites, and Porapak have been successfully identified as adsorbents. Those adsorbents could adsorb methane at atmospheric pressure and room temperature. Therefore, in this scholarly study, a new method using adsorbents to detect points of methane leakage that can cover a wide-scale area was developed. In the beginning, the most capable adsorbent should be determined by quantifying adsorbed methane amount. Furthermore, checking the possibility of adsorption in the column diffusion and desorption method of adsorbents is equally necessary. The most capable adsorbent was activated carbon (AC), which can adsorb 1.187×10^{-3} mg-CH_4/g-AC. Hereinafter, activated carbon successfully can adsorb methane through column diffusion, which simulates the situation of on-site measurement. The specific amount of adsorbed methane when the initial concentrations of CH_4 in a bag were 200 ppm, 100 ppm, and 50 ppm was found to be 0.818×10^{-3} mg-CH_4/g-AC, 0.397×10^{-3} mg-CH_4/g-AC, 0.161×10^{-3} mg-CH_4/g-AC, respectively. Desorption of activated carbon analysis shows that methane concentration increases during an hour in the temperature bath under 80 °C. In conclusion, soil methane leakage points can be detected using activated carbon by identifying the observed methane concentration increase.

Keywords: adsorption; methane leakage; column diffusion; carbon storage; performance measurement

1. Introduction

Methane (CH_4) is an important greenhouse gas, with a global warming potential (GWP) 28–36 times that of carbon dioxide (CO_2) in a time horizon of 100 years [1]. The European Environment Agency [2] has revealed that methane concentration doubled, from roughly 900 ppb to about 1800 ppb, over the past century. Coal and natural gas fields accumulated approximately 100 teragrams of atmospheric methane, which is the third-largest contribution of global methane emissions [3].

CH_4 is produced in soils by anaerobic methanogenic microbes and consumed by aerobic methanotrophic bacteria under some circumstances [4]. Studies reported that the microseepage of methane from deep oil traps is typically viewed as a local source of atmospheric methane [5,6]. Correspondingly, the subsurface leakage of methane is measured to appropriately recognize the dynamics of gas migration on the areas of abandoned shafts, mine ventilation termination, closed and remediated coal mines [7]. It also showed methane can move to the surface along the mining-induced fracture zones, which are caused by mining activities.

Recently, a static flux chamber method has been applied to study methane leakage into the atmosphere on the natural gas and coalfield, e.g., in a natural gas field in Tarim Basin, China [8], oil and gas wells [9], abandoned shafts of coal mines [10]. The flux chamber method indicatively has a range of measurements between 0.1 ppm and

50 ppm with 0.25 ppb precision [11]. However, it is only suitable for point measurements, and it consumes considerable time because many flux chamber measurements must be needed [12].

A variety of microporous materials such as activated carbon (AC) and zeolites have been studied and developed for the physical adsorption of gas mixtures containing CO_2 and CH_4 [13,14]. Methane can be adsorbed by activated carbon under low pressure 1 bar at 303 K [15]. Meanwhile, Porapak is commonly used in a packed column of gas chromatography to adsorb methane. Therefore, a new alternative method using one of those three adsorbents, which is the best adsorbent to detect points of methane leakage on the wide-scale area of coal or natural gas fields was particularly studied.

2. Results

2.1. Screening Adsorbents

The experiment objectively evaluating the most potent adsorbent was successfully conducted. In this comprehensive study, commercial activated carbon, zeolites, and Porapak, with specific characteristics and properties, were used. Before the test, using the adsorbents, the baseline experiment was essentially conducted; briefly, methane concentration was analyzed for 60 min in the Tedlar bag that contained no adsorbents in it. Moreover, three mass variations of the adsorbents were prepared. Then, 5 g of the adsorbents was applied at the first attempt, then added 5 g increments of the adsorbents for the following experiments up to 15 g.

Activated carbon exhibited the best result (Figure 1). The final concentration of methane was persistently the lowest, compared with others (Figure 1), which means that more methane was adsorbed by activated carbon. The de-escalatory concentration of methane when using porapak and zeolites was positively below the baseline, but the reduction in concentration was not extremely significant (Figure 1). All in all, the decreased concentrations of methane when using the adsorbents were consistently below the baseline (Figure 1), indicating methane was successfully adsorbed onto the surface of the adsorbents.

Based on the results above, it can be inferred that the concentration was relatively stable after 20 min. A decrease in methane concentration was observed in the first 20 min. Therefore, the amount of adsorbed CH_4 was calculated with time intervals of 20 min.

$$\text{Adsorbed CH}_4\left(\frac{\text{L}}{\text{g}}\right) = \frac{(\Delta\text{CH}_{4\ \text{concentration}}) \times \text{ Gas bag vol. (L)}}{\text{Adsorbent mass (g)}} \quad (1)$$

$$\begin{aligned}\text{Adsorbed CH}_4\left(\tfrac{\text{L}}{\text{g}}\right) &= \tfrac{(102.873-58.368)\ \text{ppm}\ \times 0.2\ \text{L}}{5\ \text{g}} \\ &= 1.780 \times 10^{-6}\ \tfrac{\text{L}-\text{CH}_4}{\text{g}-\text{AC}}\end{aligned} \quad (2)$$

At room temperature, the volume of 1 mol gas is 24 L; therefore,

$$\text{Adsorbed CH}_4\left(\frac{\text{mg}}{\text{g}}\right) = \frac{\text{Adsorbed CH}_4\left(\frac{\text{L}}{\text{g}}\right)}{24\ \text{L}} \times \text{Mr}_{\text{CH}_4}\left(\frac{\text{g}}{\text{mol}}\right) \times 1000 \quad (3)$$

$$\begin{aligned}\text{Adsorbed CH}_4\left(\tfrac{\text{mg}}{\text{g}}\right) &= \tfrac{1.780 \times 10^{-6}\left(\frac{\text{L}}{\text{g}}\right)}{24\ \text{L}} \times 16\left(\tfrac{\text{g}}{\text{mol}}\right) \times 1000 \\ &= 1.187 \times 10^{-3}\tfrac{\text{mg}-\text{CH}_4}{\text{g}-\text{AC}}\end{aligned} \quad (4)$$

Activated carbon had the highest value of adsorbed methane volume by 1.187×10^{-3} mg-CH_4/g-AC, followed by Porapak by 0.150×10^{-3} mg-CH_4/g-P and zeolites by 0.017×10^{-3} mg-CH_4/g-Z when 5 g of adsorbents were optionally used (Table 1). Nevertheless, the value amount of adsorbed CH_4 of activated carbon degraded by proportionately increasing activated carbon mass, whereas zeolites demonstrated the opposite. The value of adsorbed CH_4 was assumed to be the same. This discovered phenomenon is discussed in the next section.

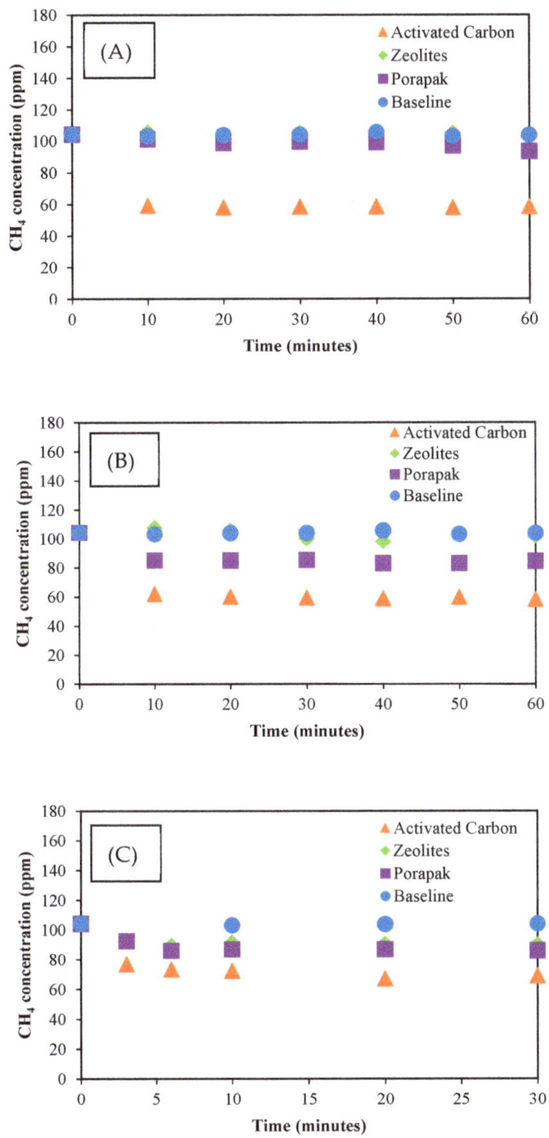

Figure 1. Adsorption performance of adsorbents: (**A**) 5 g; (**B**) 10 g; (**C**) 15 g.

Table 1. Amount of adsorbed CH_4.

Adsorbent	**Amount of Adsorbed CH_4 (10^{-3} mg/g Adsorbent)**		
	5 g	**10 g**	**15 g**
Activated carbon	1.187	0.585	0.317
Zeolites	0.017	0.044	0.117
Porapak	0.150	0.175	0.155

Based on the results, a graph was generated to investigate the gradient of the methane concentration reduction line of each adsorbent, which varied by the determined mass of the adsorbent (Figure 2). Activated carbon displayed the lowest gradient, followed by Porapak and zeolites, even though the performance of the activated carbon decreased by gradually adding 5 g of it (Figure 2). This sufficiently confirmed the previous results (Table 1).

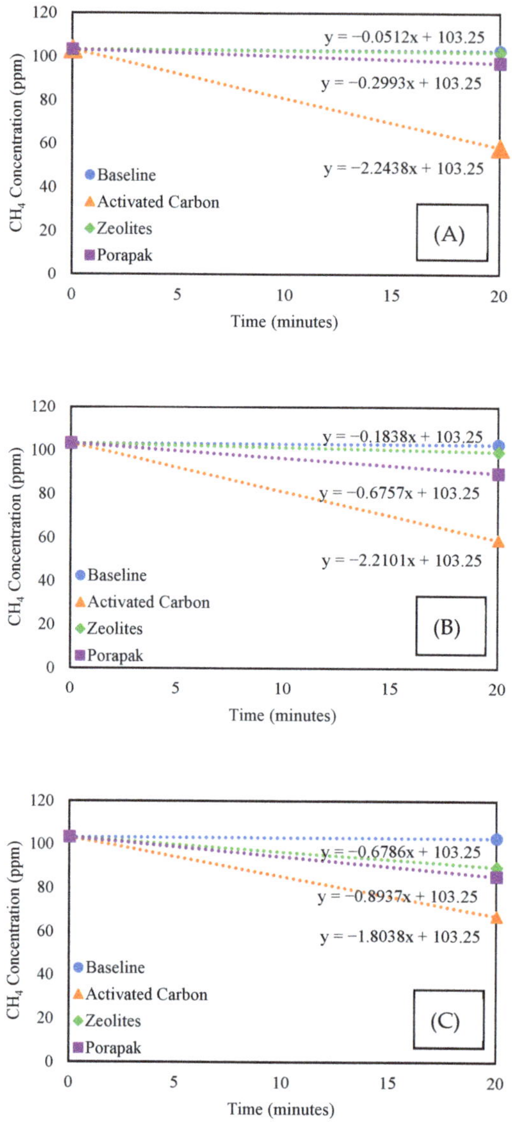

Figure 2. Graph for the 20 min adsorption analysis: (**A**) 5 g, (**B**) 10 g, and (**C**) 15 g adsorbents.

As can be seen in Table 1, the value of the volume of adsorbed CH_4 by the activated carbon became lower by adding more 5 g increments of activated carbon. Contrastingly, the value of the volume of adsorbed CH_4 by the zeolites became higher by adding more 5 g increments of zeolites. Porapak mass analytically had no significant effect on the volume of the amount of adsorbed CH_4. To understand and interpret the underlying reasons behind these outcomes, further examination was carried out.

As previously mentioned, the same materials from the previous experiment were used, then added 5 g for the next examination. Identical experiments using the same adsorbent for the first and second measurements were then necessarily performed. The moderate changes in CH_4 concentration were tracked for 20 min. This investigation was conducted as a particular reference for further experiments to investigate whether it is using the adsorbent more than once would become problematic.

As evident in the gradient of the methane concentration reduction line, the first and second measurements sufficiently reveal that activated carbon strongly decreased in terms of its adsorption performance when used for the second measurement (Figure 3A). The gradient of the second measurement tended to have a similar slope to the first, indicating there was no more methane adsorbed in the second measurement than in the first. All in all, fresh activated carbon could remarkably be used in every single subsequent experiment.

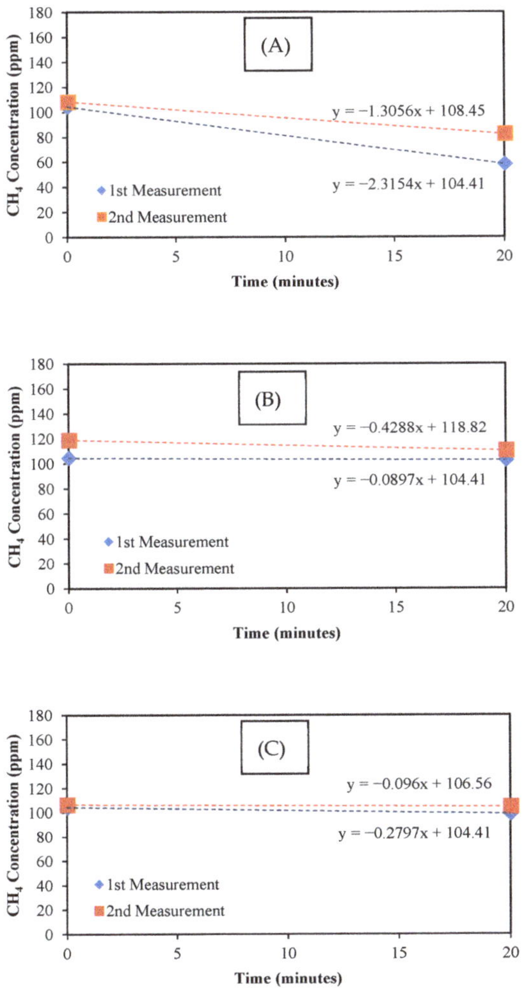

Figure 3. The first vs. second measurements of 5 g additions of (**A**) activated carbon, (**B**) zeolites, and (**C**) Porapak.

Zeolites and Porapak showed stable performance for the first and second measurements. That is why for every documented increase in the mass of the adsorbent, the value of the amount of adsorbed CH_4 per gram of adsorbent quantitatively was almost the same (Table 1).

2.2. Column Diffusion

Examining the possibility of adsorption in the gas column diffusion was essential to understand whether this suggested method is applicable or not. The experiments were conducted by injecting no adsorbent inside the column to properly examine the natural diffusion behavior of methane in the column from one Tedlar bag to another Tedlar bag. The methane concentrations of both Tedlar bags as the baseline, marked by the blue color, were carried out. The red color illustrates the methane concentration of both Tedlar bags using activated carbon in the column.

Three types of measurements were conducted. The sets of initial concentrations of methane in the Tedlar bag A—200 ppm, 100 ppm, and 50 ppm—were performed to observe the column diffusion phenomenon. Those three initial concentrations were implemented for easier observation of the diffusion process. If it was excessively low, the diffusion process in the column would be difficult to observe. As regards the time limit for when the test was completed, methane concentration in the column diffusion was equitably evaluated for 4 h because methane concentration does not fluctuate significantly after 4 h, indicating the diffusion of methane from one Tedlar bag to another Tedlar bag through the column naturally lasts 4 h.

From the results of the laboratory experiments, the final methane concentration when activated carbon was present in the column was always lower than the final methane concentration when no adsorbent was in the column (Figure 4). This means that methane flowing abundantly from Tedlar bag A to Tedlar bag B was successfully adsorbed. However, the smaller the initial value of the CH_4 concentration in the first Tedlar bag was, the smaller the difference in value between the red and blue dots in the second Tedlar bag would become because the measurable amount of methane in the analyzed system became more insignificant.

Adsorbed methane amount on this completed experiment was relatively the same as the screening results of the experiment using adsorbents (Table 1), prominently representing that the activated carbon performance appeared to be constant even though it was a separate type of experiment. However, the lower the initial concentration was, the lesser was the adsorbed methane amounts per gram of activated carbon (Table 2) because the amount of methane in the designed system became gradually smaller and more limited so that the methane was increasingly difficult to adsorb onto the activated carbon surface. The amount of adsorbed methane when the initial concentrations of CH_4 in a bag were 200 ppm, 100 ppm, and 50 ppm was found to be 0.818×10^{-3} mg-CH_4/g-AC, 0.397×10^{-3} mg-CH_4/g-AC, 0.161×10^{-3} mg-CH_4/g-AC, respectively.

Table 2. Amount of CH_4 adsorption in a column.

Initial Concentration (ppm)	Final Concentration (ppm)		Adsorbed CH_4 Amount (10^{-3} mg/g-AC)
	Without AC	With AC	
200	79.7	28.2	0.818
100	38.2	13.2	0397
50	18.4	8.24	0.161

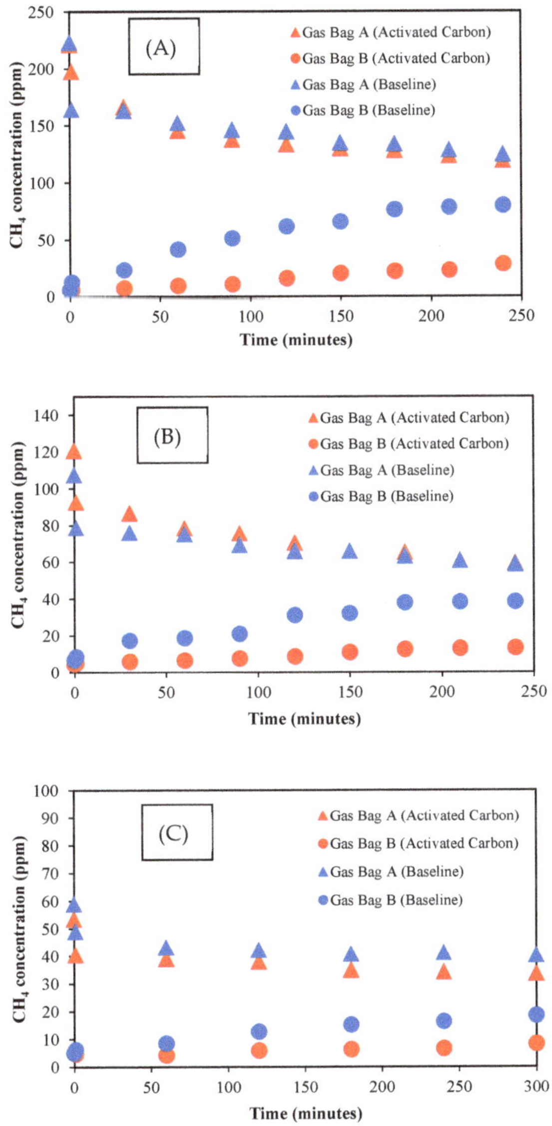

Figure 4. Column CH_4 adsorption of (**A**) 200 ppm, (**B**) 100 ppm, and (**C**) 50 ppm initial concentration.

3. Discussion

3.1. Thermal Desorption

Same as the previous experiment, conducting the baseline procedure was vital for validating the experimental results. Baseline data indicate the thermal desorption data of the fresh activated carbon that had not undergone an adsorption process. The increase in methane concentration had to be checked because methane basically must be present in activated carbon. Results sufficiently reveal there was a slight increase in the concentration of methane at the baseline experiment, indicating that methane was present on the fresh activated carbon surface before desorption (Figure 5).

Five sets of variations of the initial concentration in one Tedlar bag before adsorption were planned to be tested. Based on preference, the investigation began with the highest one, which was 100 ppm, then 50 ppm, 25 ppm, 10 ppm, and 5 ppm. The increase in the concentration of methane was precisely measured during an hour of the thermal desorption process from each of these initial system conditions. The results show the value of the increase in the concentration of methane was always greater than the baseline experiment (Figure 5).

As an example, when the initial concentration of methane was 5 ppm, meaning there was gas leakage from one Tedlar bag to another Tedlar bag, becoming ideally 2.5 ppm and 2.5 ppm, methane could be detected by indicating the concentration difference or the increase in the methane concentration of 5 ppm measurement, which was still higher than the concentration difference of the baseline experiment. To reasonably conclude, 5 ppm leakage point of methane from the subsurface to the atmosphere could be correctly detected using activated carbon columns.

In terms of simulating the detection of methane leak points from the subsurface to the atmosphere, a column was prepared in which there was activated carbon, with the opening of one column connected by a 50 L Tedlar bag, and the other openings were directly connected to the open air (atmosphere). An initial concentration value of 5 ppm was set because the atmospheric methane concentration averaged 1.9 ppm so that methane could be diffused from the Tedlar bag into the atmosphere. The methane concentration difference from thermal desorption process of the 50 L Tedlar bag with 5 ppm CH_4 was significantly higher than the methane concentration difference in the baseline measurement, indicating that activated carbon could be appropriately applied as an indicator and instrument to detect methane leakage point (Figure 5).

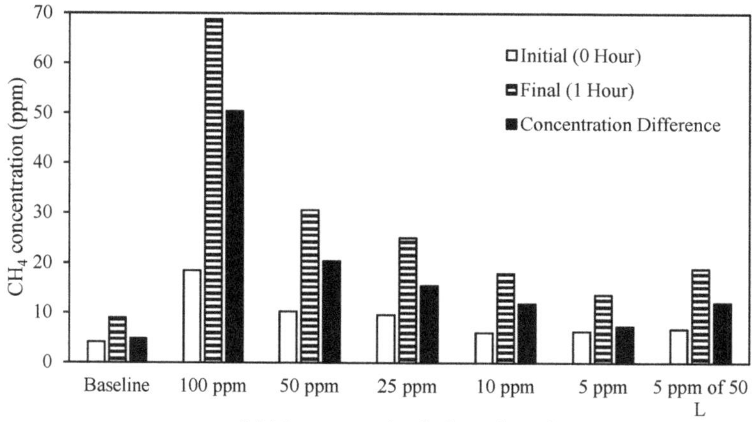

Figure 5. CH_4 thermal desorption with variations of initial concentration before adsorption.

3.2. *Low-Cost Method for Monitoring Methane Leakage*

The tests carried out were considerably influential for determining the time parameters and the design of the proposed methods (Figure 6). Activated carbon columns had to be installed into the soil layer. Adsorbents needed 5 h for adsorbing methane. Then, the columns were gathered, brought to the lab, and placed into the oven, i.e., constant temperature bath, for thermal desorption. The desorption process required 1 h. Eventually, an increase in methane concentration before and after the columns were cautiously placed in the oven was accurately analyzed by gas chromatography. The value of methane leakage was estimated after 6 h of tests.

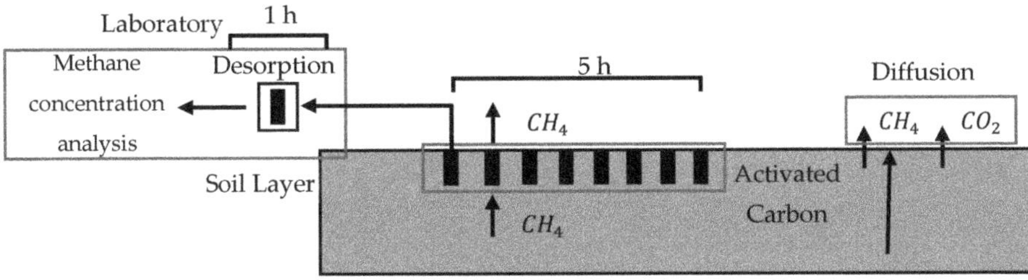

Figure 6. Methane leakage monitoring conceptual design of proposed method.

4. Materials and Methods

Methane concentrations in the gas samples were accurately analyzed in the Laboratory of Resources Production and Safety Engineering, Kyushu University, Japan, using gas chromatography (GC-14B, Shimadzu, Japan) with a flame ionization detector (FID) that was calibrated by the methane standard gas of (99.0%) before each of measurements. Gas chromatography had to be calibrated to construct the calibration curves. Before performing the measurements, methane standard gas was injected into gas chromatography with the volume of 0.02 mL, 0.015 mL, 0.0125 mL, 0.01 mL, 0.0075 mL, 0.005 mL, and 0.0025 mL. Afterward, linear calibration curves of methane volume in relation to the area of the graph showing methane peaks in gas chromatography were generated. The linear equation of the curves was then used for determining the methane concentration of the measurements. The working conditions of the chromatographer were set as follows: a GC stainless column 6.0 m × 3.0 mm I.D. (packed column) that was inserted by 80 or 100 mesh Porapak Q; argon (Ar^{18}) carrier gas; column temperature of 100 °C; injection temperature of 100 °C; an FID detector with a constant temperature of 100 °C. Methane flowed simultaneously with a carrier gas (Ar^{18}) in the orbicular column. Methane was introduced into a hydrogen flame inside the FID detector. Methane in the selected sample would produce ions when burnt, whereas the carrier gas would not. Ions were detected ordinarily by employing a metal collector biased with a high DC voltage. Lastly, readings of methane concentration were processed with the "CDS" (ver. 2.83) software.

4.1. Adsorbents

There are fundamentally two activation methods to generate activated carbons: physical and chemical. They can be used separately or in a conceivable combination with one another. On the other hand, commercial activated carbon is widely available in the market. Thus, commercial activated carbon manufactured by Junsei Chemical Co. Ltd., which has a granular form, was preferentially used in this study (Figure 7). Moreover, synthetic beads zeolites with a specified size of 1.40–2.36 mm (8–12 mesh) fabricated by Fujifilm Wako Pure Chemical Corporation and Porapak type Q with a size of 80–100 mesh, the chemical name of which is poly (4-ethyl styrene-CO-divinylbenzene), produced by Waters Corporation, were used in this study (Figure 7). The value of the specific surface area of activated carbon, zeolites, Porapak calculated from the characteristics data of commercial products is 1000 m^2/g, 500–600 m^2/g, and 600 m^2/g, respectively. In addition, information of the bulk density of activated carbon, zeolites, and Porapak is 0.46 g/cm^3, 0.70–0.82 g/cm^3, and 0.34 g/cm^3, respectively. Further, 5 g, 10 g, and 15 g of each adsorbent were provided to evaluate which adsorbent is the best in terms of the possibility of diffusion in the column and thermal desorption.

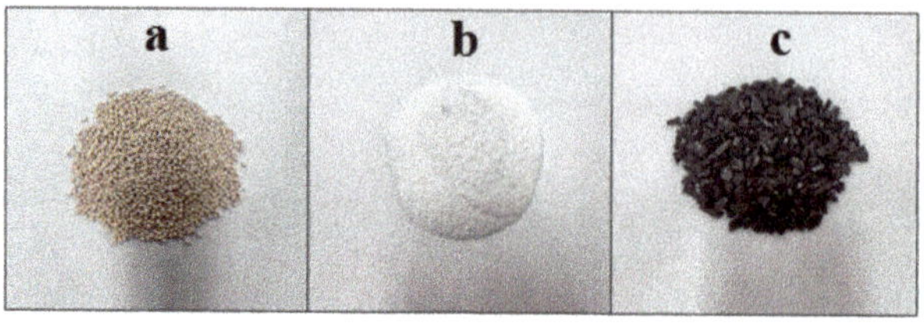

Figure 7. Adsorbents: zeolites (**a**), porapak (**b**), and activated carbon (**c**).

4.2. Screening Adsorbents

The first subject of our study was evaluating the most potent adsorbent. To start with, the adsorbent was carefully inserted into the 1 L Tedlar gas bag, and the Tedlar bag was made in a vacuum condition. The funnel was used to facilitate the entry of the adsorbent into the opening of the Tedlar bag, as it was arduous to insert the adsorbent directly, considering the opening was relatively narrow. When the adsorbent grain size was overly large, such as with activated carbon, it was then ground using mortar and stamper to make the grain size smaller. Thereafter, 200 mL of air and 0.02 mL of methane were reciprocally injected. This analytically means the methane concentration in the Tedlar bag was 100 ppm. Lastly, the sample gas was extracted every 10 min to be accurately analyzed the methane concentration using gas chromatography (Figure 8).

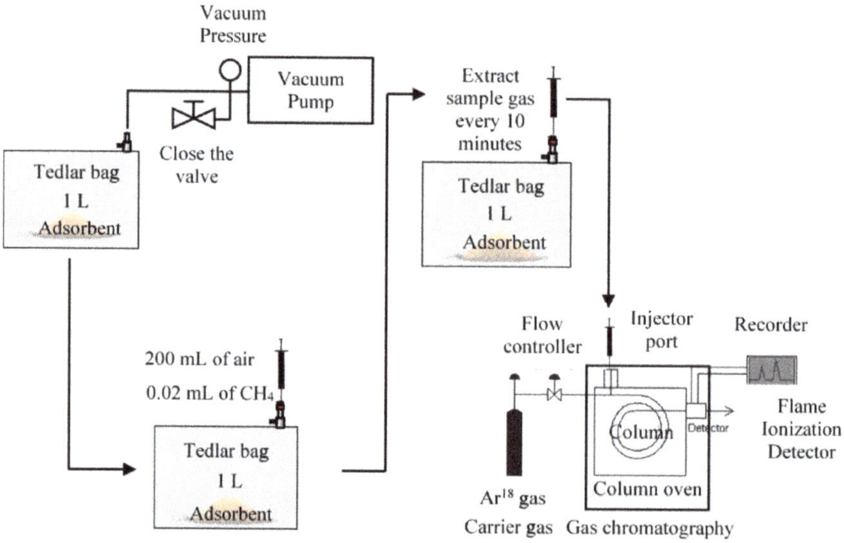

Figure 8. Experimental setup of screening the adsorbents.

The experiments were conducted by varying the mass of each adsorbent 5 g, 10 g, and 15 g in the 1 L Tedlar bag. The first measurement was performed using 5 g of adsorbent in the Tedlar bag, then analyzed the adsorbed methane value of 5 g adsorbents. Afterward, 5 g of extra adsorbent was added into the Tedlar bag. Therefore, the second measurement was performed using 10 g of adsorbent, and the adsorbed methane value of 10 g adsorbents was then analyzed as well. Next, another 5 g increment of adsorbent was added, amounting to 15 g of adsorbent in the Tedlar bag for the subsequent measurement to analyze the adsorbed methane value. The weights were conscientiously evaluated by laboratory LCD analytical digital balance. Briefly, the same materials from the previous measurement were used, then added 5 g for the subsequent measurement. Therefore, the adsorption performance analysis of the adsorbents that were used more than once needed to be interpreted. This procedure can appropriately provide reasonable insights for researchers who study the application of the adsorbents more than once by conducting experiments to compare the first vs. second adsorption measurements of adsorbents.

Adsorbed methane amount can be measured by following the fundamental equation below, which considers the concentration difference of methane ($\Delta CH_{4\ conc.}$) between initial concentration and final concentration [16]:

$$\text{Adsorbed } CH_4\left(\frac{L}{g}\right) = \frac{(\Delta CH_{4\ \text{concentration}}) \times \text{Gas bag vol. (L)}}{\text{Adsorbent mass (g)}} \quad (5)$$

At room temperature, the volume of 1 mol gas is 24 L; therefore,

$$\text{Adsorbed } CH_4\left(\frac{mg}{g}\right) = \frac{\text{Adsorbed } CH_4\left(\frac{L}{g}\right)}{24\text{ L}} \times Mr_{CH_4}\left(\frac{g}{mol}\right) \times 1000 \quad (6)$$

4.3. Column Diffusion

The second subject of the study was to examine the possibility of adsorption in the gas column diffusion. Two Tedlar bags of 1 L were prepared in vacuum condition. The first Tedlar bag (B) was only injected with 500 mL of air. The other (A) was injected with 500 mL of air and methane with decided variations of volume. At the start of the procedure, 0.1 mL of methane was injected, which means 200 ppm of methane was inside the Tedlar bag. Then, 100 ppm and 50 ppm of the methane in the Tedlar bag were prepared to be analyzed as the source of methane (Figure 9). The diffusion occurred spontaneously, after which the gas was extracted from each of the Tedlar bags. After this step, the methane concentration was properly analyzed by gas chromatography throughout 5 h measurements. Theoretically, methane will flow from the higher concentration to the lower concentration. The column acted as a medium for the diffusion of methane. The purpose of the experiment was primarily to observe how the diffusion of methane can spontaneously occur in the presence and absence of the adsorbent in the column.

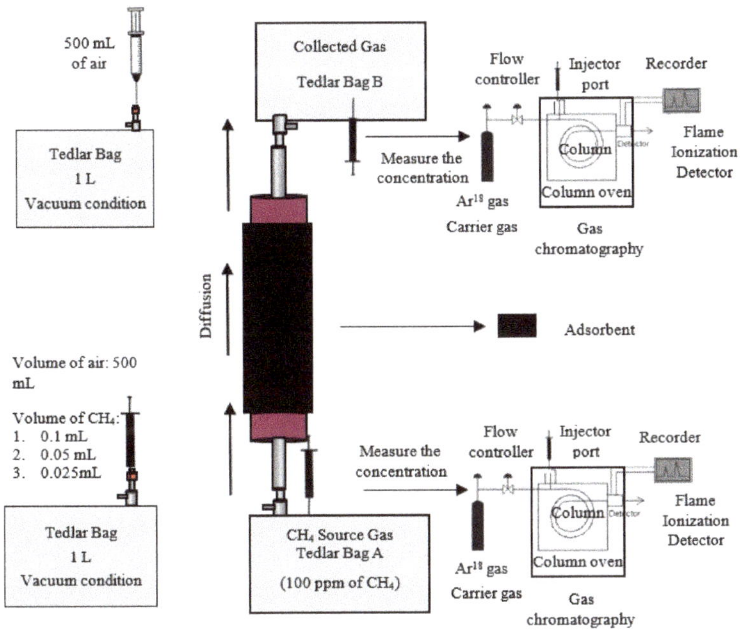

Figure 9. Experimental setup of adsorption in the column.

Concerning the main objective, which was to develop a method to detect methane leakage points from the subsurface to the atmosphere by using adsorbent, the laboratory experimental was designated to simulate methane leakage to the atmosphere detected by using a column. To this end, 50 L of Tedlar gas bag containing 5 ppm of methane from the injection of 50 L of air and 0.25 mL of methane was arranged similarly as a source of methane from the subsurface (Figure 10). The column in this experiment simulated an adsorbent column being placed on the ground. Another opening of the column was left open because it simulated the dynamic flow of methane from the subsurface to the open air in the atmosphere through the ground.

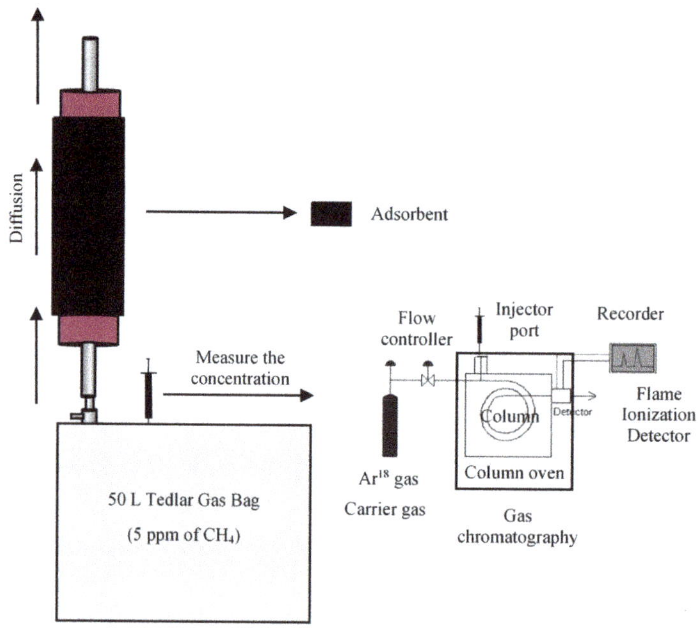

Figure 10. Experimental setup of diffusion to the atmosphere.

4.4. Thermal Desorption

Thermal Desorption of adsorbate from activated carbon at 80–120 °C under atmospheric pressure has been experimentally studied [17]. For this study, the closed column was cautiously placed into the constant temperature bath under 80 °C, and the sample gas inside the column was extracted into a syringe before and after desorption (Figure 11).

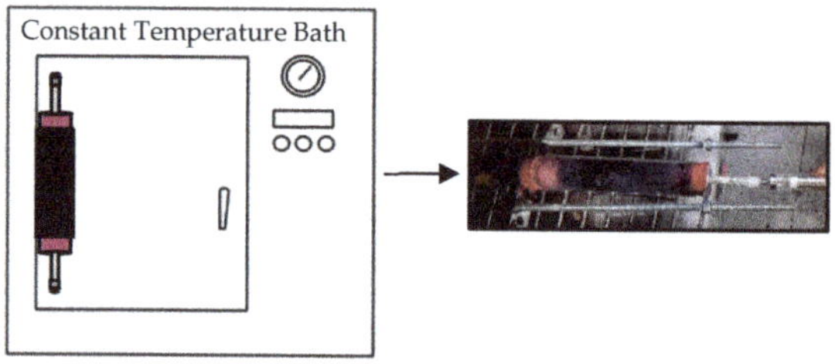

Figure 11. Thermal desorption.

This experiment was directly related to the column diffusion experiment. If the column diffusion experiments were successfully conducted and exhibited acceptable results indicating adsorption by the adsorbent, they were repeated using the lower initial concentration of methane. After diffusion occurred through the column, which required approximately five hours, both column openings had to be sealed to ensure methane would not leak from the column. The gas was extracted from the column and analyzed by gas chromatography with the initial concentration of methane inside the column before adsorption. The column was delicately placed in a constant temperature bath, then the gas was extracted from the column after an hour in the constant temperature bath for analysis by gas chromatography.

4.5. Proposed Method

The above catena of methodical processes (Sections 4.2–4.4) was practically applied as a single flow for developing a new method (Figure 12). Each process was first tested to reliably determine whether the proposed method could be implemented effectively for monitoring methane leakage. In this proposed method, columns were placed randomly following the field conditions according to which the methane leakage would be estimated. Those columns were charged by adsorbent and replaced with new columns every prescribed period. The columns that were recovered from the field were carried to the laboratory. The gas adsorbed on the adsorbent was desorbed under high-temperature conditions. Under those circumstances, methane concentration was correctly analyzed using a detector such as a flame ionization detector (FID) in the gas chromatograph. After a considerable time, methane leakage in the field could be estimated.

Figure 12. Schematic process of proposed method.

5. Conclusions

This study developed a low-cost method to detect methane leakage points from the subsurface to the atmosphere through the ground. Activated carbon demonstrated the highest value of adsorbed methane amount, at 1.187×10^{-3} mg-CH_4/g-AC, followed by Porapak, at 0.150×10^{-3} mg-CH_4/g-P, and zeolites at 0.017×10^{-3} mg-CH_4/g-Z when 5 g of adsorbents were used. Therefore, activated carbon was the best adsorbent by screening the adsorption capacity of those three adsorbents. In addition, activated carbon could adsorb methane during the diffusion process between two Tedlar bags, which had different methane concentrations in column experiments. The detectable amount of adsorbed methane when the initial concentrations of CH_4 in a bag were 200 ppm, 100 ppm, and 50 ppm was found to be 0.818×10^{-3} mg-CH_4/g-AC, 0.397×10^{-3} mg-CH_4/g-AC, 0.161×10^{-3} mg-CH_4/g-AC, respectively, indicating that its performance appeared to be constant even though it was a different type of experiment. The analytical results reveal that the concentration difference of methane when 5 ppm of the initial concentration of methane adsorption was used with activated carbon was higher than the concentration difference of the baseline on thermal desorption, which indicates that activated carbon can detect methane leak points from subsurface to the atmosphere even with an exceedingly small concentration of 5 ppm as a source of methane. The proposed method has a limitation—it merely indicates the point of methane leakage. The value of methane emissions in the overall field is not able to be measured appropriately. However, this method can precisely indicate the points of methane leakage up to 5 ppm leakage from the subsurface.

Author Contributions: Conceptualization, Y.S. and M.A.F.; methodology, Y.S.; software, Y.S.; validation, M.A.F., Y.S., N.P.W. and S.S.; formal analysis, M.A.F.; investigation, Y.S.; resources, N.P.W. and S.S.; data curation, M.A.F.; writing—original draft preparation, M.A.F.; writing—review and editing, M.A.F.; visualization, M.A.F.; supervision, Y.S. and N.P.W.; project administration, N.P.W. and S.S.; funding acquisition, Y.S. and S.S. All authors have read and agreed to the published version of the manuscript.

Funding: This research received no external funding. The APC was funded by Yuichi Sugai.

Institutional Review Board Statement: Not applicable.

Informed Consent Statement: Not applicable.

Conflicts of Interest: The authors declare no conflict of interest. The funders had no role in the design of the study; in the collection, analyses, or interpretation of data; in the writing of the manuscript, or in the decision to publish the results.

References

1. Intergovernmental Panel on Climate Change (IPCC). *Mitigation of Climate Change: Energy Systems*; Fifth Assessment, Report; IPCC: Cambridge, UK; Cambridge University Press: Cambridge, UK, 2014.
2. European Environment Agency (EEA). EEA air pollutant emission inventory guidebook-2009. In *Technical Guidance to Prepare National Emission Inventories*; EEA Technical Report No 6/2009; EEA: Copenhagen, Denmark, 2009.
3. Maasakkers, J.D.; Jacob, D.J.; Sulprizio, M.P. Global distribution of methane emissions, emission trends, and OH concentrations and trends inferred from an inversion of GOSAT satellite data for 2010–2015. *Atmos. Chem. Phys.* **2019**, *19*, 7859–7881. [CrossRef]
4. Le Mer, J.; Roger, P. Production, oxidation, emission, and consumption of methane by soils: A review. *Eur. J. Soil Biol.* **2001**, *37*, 25–50. [CrossRef]
5. Klusman, R.W.; Jakel, M.E. Natural microseepage of methane to the atmosphere from the Denver-Julesburg basin, Colorado. *J. Geophys. Res. Atmos.* **1998**, *103*, 28041–28045. [CrossRef]
6. Mosier, A.R.; Schimel, D.S. Influence of agricultural nitrogen on atmospheric methane and nitrous oxide. *Chem. Ind.* **1991**, *23*, 874–877.
7. Sechman, H.; Kotarba, M.J.; Fiszer, J.; Dzieniewicz, M. Distribution of methane and carbon dioxide concentrations in the near-surface zone and their genetic characterization at the abandoned "Nowa Ruda" coal mine (lower Silesian Coal Basin, SW Poland). *Int. J. Coal Geol.* **2013**, *116*, 1–16. [CrossRef]
8. Tang, J.; Xu, Y.; Wang, G.; Etiope, G.; Han, W.; Yao, Z.; Huang, J. Microseepage of methane to the atmosphere from the Dawanqi oil-gas field, Tarim Basin, China. *J. Geophys. Res. Atm.* **2017**, *122*, 4353–4363. [CrossRef]
9. Riddick, S.N.; Denise, L.; Mauzerall, M.A.; Celia, M.K.; Bressler, K.; Chu, C.; Gum, C.D. Measuring methane emissions from abandoned and active oil and gas wells in West Virginia. *Sci. Total Environ.* **2019**, *651*, 1849–1856. [CrossRef] [PubMed]
10. Sechman, H.; Kotarba, M.J.; Kedzior, S.; Dzieniewicz, M.; Romanowski, T.; Twaróg, A. Distribution of methane and carbon dioxide concentrations in the near-surface zone, genetic implications, and evaluation of gas flux around abandoned shafts in the Jastrzębie-Pszczyna area. *Int. J. Coal Geol.* **2019**, *204*, 51–69. [CrossRef]
11. Xu, L.; Vath, R.; McCoy, J. *Chamber-Based Soil Methane Flux Measurement*; LI-COR Biosciences: Lincoln, NE, USA, 2018.
12. Fredenslund, A.M.; Scheutz, C.; Kjeldsen, P. Tracer method to measure landfill gas emissions from leachate collection systems. *Waste Manag.* **2010**, *30*, 2146–2152. [CrossRef] [PubMed]
13. Dantas, T.L.P.; Luna, F.M.T.; Silva, I.J., Jr.; Torres, A.E.B.; de Azevedo, D.C.S.; Rodrigues, A.E.; Moreira, R.F.P.M. Modelling of the fixed-bed adsorption of carbon dioxide and a carbon dioxide-nitrogen mixture on zeolite 13X. *Braz. J. Chem. Eng.* **2011**, *28*, 533–544. [CrossRef]
14. Belmabkhout, Y.; Sayari, A. Adsorption of CO_2 from dry gases on MCM-41 silica at ambient temperature and high pressure. 2: Adsorption of CO_2/N_2, CO_2/CH_4 and CO_2/H_2 binary mixtures. *Chem. Eng. Sci.* **2009**, *64*, 3729–3735. [CrossRef]
15. Rios, R.B.; Silva, F.W.M.; Torres, A.E.B.; Azevedo, D.C.; Cavalcante, C.L. Adsorption of methane in activated carbons obtained from coconut shells using H_3PO_4 chemical activation. *Adsorption* **2009**, *15*, 271–277. [CrossRef]
16. Abdeen, Z.; Somaia; Mohammad, G. Study of the Adsorption Efficiency of an Eco-Friendly Carbohydrate Polymer for Contaminated Aqueous Solution by Organophosphorus Pesticide. *J. Org. Polym. Mater.* **2014**, *4*, 16–28. [CrossRef]
17. Ion, V.A.; Parvulescu, O.C.; Dobre, T.; Duteanu, N.; Nitu, S.G. Modelling of Thermal Desorption of Volatile Organic Compounds from Activated Carbon. *Rev. De Chim.* **2015**, *66*, 703–706.

methane

MDPI

Article

Comparison of Methods to Segment Variable-Contrast XCT Images of Methane-Bearing Sand Using U-Nets Trained on Single Dataset Sub-Volumes

Fernando J. Alvarez-Borges [1,2], Oliver N. F. King [2,*], Bangalore N. Madhusudhan [1], Thomas Connolley [2], Mark Basham [3] and Sharif I. Ahmed [2]

1 Faculty of Engineering and Physical Sciences, University of Southampton, Southampton SO17 1BJ, UK
2 Diamond Light Source Ltd., Harwell Science & Innovation Campus, Didcot OX11 0DE, UK
3 The Rosalind Franklin Institute, Harwell Science & Innovation Campus, Didcot OX11 0QS, UK
* Correspondence: olly.king@diamond.ac.uk

Abstract: Methane (CH_4) hydrate dissociation and CH_4 release are potential geohazards currently investigated using X-ray computed tomography (XCT). Image segmentation is an important data processing step for this type of research. However, it is often time consuming, computing resource-intensive, operator-dependent, and tailored for each XCT dataset due to differences in greyscale contrast. In this paper, an investigation is carried out using U-Nets, a class of Convolutional Neural Network, to segment synchrotron XCT images of CH_4-bearing sand during hydrate formation, and extract porosity and CH_4 gas saturation. Three U-Net deployments previously untried for this task are assessed: (1) a bespoke 3D hierarchical method, (2) a 2D multi-label, multi-axis method and (3) RootPainter, a 2D U-Net application with interactive corrections. U-Nets are trained using small, targeted hand-annotated datasets to reduce operator time. It was found that the segmentation accuracy of all three methods surpass mainstream watershed and thresholding techniques. Accuracy slightly reduces in low-contrast data, which affects volume fraction measurements, but errors are small compared with gravimetric methods. Moreover, U-Net models trained on low-contrast images can be used to segment higher-contrast datasets, without further training. This demonstrates model portability, which can expedite the segmentation of large datasets over short timespans.

Keywords: U-Net; methane hydrates; microtomography; sediment microstructure; semantic segmentation

1. Introduction

Deep sea sediments and perfmafrost host large quantities of methane (CH_4), an energy source and potent greenhouse gas that may be a contributor to climate change [1,2]. Much of this CH_4 is present as hydrates (clathrates), that is, solid crystalline lattices of water at low temperature and high pressures that enclose CH_4 molecules. 164 m^3 of CH_4 gas at normal temperature and pressure can be stored in one m^3 of hydrate [3]. However, the extent of the world-wide CH_4 hydrate inventory is subject to considerable uncertainty [4,5]. This is in part due to discrepancies between measurements produced by geophysical and electrical resistivity methods [6,7]. These discrepancies are potentially associated with hydrate and CH_4 gas distribution heterogeneity in the host soil [8]. Uncertainties regarding the global CH_4 hydrate inventory affect resource estimation and CH_4 emission prediction models [5,9,10]. CH_4 hydrate formation and dissociation has also been associated with changes in the mechanical characteristics of the host sediment. For instance, hydrates may strengthen and stiffen the sediment matrix by creating inter-grain cementation bonds [11, 12]. This is speculated to lead to, for example, underwater slides that may trigger tsunami or damage seabed infrastructure such as cables and pipelines [13–15].

Recently, researchers have shown that X-ray computed tomography (XCT) can be used to successfully detect hydrate and CH_4 gas bubble distribution heterogeneity and characterise changes in sediment microstructure associated with hydrate formation and dissociation [8,16–18]. This has been possible in great part due to advancements in image segmentation techniques. Segmentation is the process of classifying 2D pixels or 3D voxels into regions, for example, the solids, liquids and gases present in XCT images of geomaterials. Microstructural parameters such as grain, pore and bubble size, shape and orientation can then be derived from the segmented image, as well as volumetric (bulk) quantities such as porosity and CH_4 gas saturation ratios.

Some of the most common segmentation techniques used in geomechanics and geoscience are greyscale thresholding and watershed algorithms [19,20]. The former involves the selection of a greyscale range to classify pixels or voxels into regions of interest. Watershed algorithms redefine the image as a map where greyscale intensities form topographical elevations and catchment basins. Pixel/voxel markers within these basins are used to define the materials (or 'labels') present in the image, and the algorithm then morphologically dilates these markers until they fill their catchment basins [21,22]. Greyscale range determination in the case of thresholding techniques and marker grey value and location in the case of watershed techniques are operator and/or method dependent [19,23,24]. The values assigned to these parameters also depend on the recorded greyscale contrast, which is highly reliant on the X-ray imaging instrument and how it is optimised [25]. Sample heterogeneity or density changes during an in situ experiment will further introduce contrast variability in space and time [19,26]. As a result, thresholding and watershed segmentation are typically optimised per XCT scan, and objective comparison is difficult given that the data treatment varies between datasets. These issues often result in segmentation procedures in geomechanics and geoscience that are highly demanding of computing resources and operator time.

Novel alternative approaches have employed machine learning to segment multiple material phases present in XCT images of soil and rock samples [27,28]. For these applications, segmentations are produced via a mathematical model optimised or 'trained' using a series of 'ground truth' example segmentations of XCT images provided by the user. Within the realm of machine learning, convolutional neural networks (CNNs) are a class of deep neural networks that employ convolutional layers where the filters ('kernels') used to separate image features are learned [29]. Researchers have recently begun exploring the application of CNNs to segment XCT images of soil and rock [30–33].

U-Nets are a class of CNN originally designed to segment biomedical images [34]. The U-Net architecture is composed of downsampling (encoding/contracting) and upsampling (decoding/expanding) paths. The former reduces the spatial dimensions of the data while increasing feature information; the latter recombines spatial and feature data to generate the label image. Both paths are linked by connections that can feed the output from the contracting path directly into the corresponding level of the expanding path, thereby allowing the transfer of spatial information and the preservation of fine-grained details in the output label image. A limitation to the implementation of U-Nets (and CNNs in general) to segment XCT images of soil and rock is the preparation of training and validation datasets, which often require labour-intensive manual segmentation (hand-annotation) of many images.

The suitability of such time- and resource-saving approaches have not been examined before in research on methane-bearing sediments. This paper explores this application using three U-Net implementation strategies to segment a large number of synchrotron radiation XCT (SXCT) volumes of CH_4-bearing sand during hydrate formation and dissociation experiments; two of these U-Net implementations are entirely novel to the field of geomechanics and geoscience. These strategies were focused on the segmentation of the CH_4 phase, as it exhibited low contrast with regard to the brine-hydrate phase and was less common in the data compared to the other material phases, as shown in Figure 1a. This rendered the use of conventional thresholding or watershed techniques largely unsuitable. The aim of this investigation was thus to determine if U-Nets can accurately segment XCT images of soil samples with varying greyscale contrast between material phases using only a small number of training and validation images, therefore reducing operator and computing time and allowing for objective data comparison.

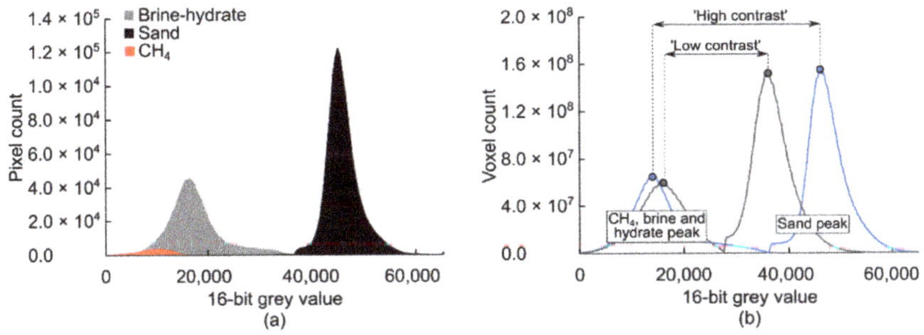

Figure 1. Grey value histograms of reconstructed and post-processed SXCT images: (**a**) of XY slice 1050 of an intermediate contrast scan, showing the frequency distribution of pixels for each material; (**b**) of two whole 3D images showing the grey value difference between histogram peaks as a measure of image contrast.

Section 2 of this paper presents the details of the in situ hydrate formation experiment as well as the methodology used to acquire, reconstruct and post-process the SXCT data. Section 3 describes the three U-Net approaches used, as well as the conventional segmentation methods and accuracy metrics used for comparison and performance analysis purposes. Results are presented and discussed in Section 3 and outcomes are summarised in Section 4.

2. Materials and Methods

2.1. Methane Gas Hydrate Formation and Dissociation Experiments

A custom rig designed and manufactured by Sahoo et al. [8] for in situ SXCT imaging of gas hydrate formation and dissociation was used in the present study. The rig is made of polyether ether ketone (PEEK) and consists of a monolithic 2 mm internal diameter by 23 mm tall cylindrical vessel with 0.8 mm thick walls and an enlarged base, as shown in Figure 2. The soil sample is placed through the bottom of the rig. The pore fluid injection pipe is connected to this inlet, as depicted in Figure 2. The rig features thermocouples at the base of the scan zone, shown in this Figure, to measure sample temperature. The SXCT imaging zone in this study corresponds to a vertically centred 1.755 mm-tall region within the 10 mm-tall scan zone.

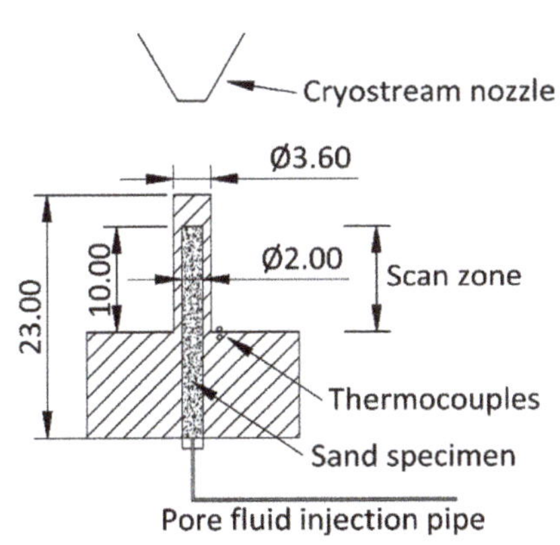

Figure 2. Cross-section sketch of hydrate test rig. Monolithic PEEK element denoted by hatched area. All units mm.

Leighton Buzzard sand Fraction E (LBE) with mean grain diameter of 100 μm was used as surrogate marine sediment. LBE is an angular silica sand widely used as standard laboratory material in geomechanics research. The sand was tamped into PEEK vessel to a target porosity of 35%. A vacuum pressure of less than 1 Pa was applied through the injection pipe to reduce air presence in the pore space. A calculated volume of brine solution (3.5% NaCl by weight, representative of deep ocean water; [35]) was thereafter injected into the sample, such that approximately 90% of the pore volume became saturated. CH_4 gas was then injected at 10 MPa and the valve to the sample closed. The sample was gradually cooled to a target constant temperature of 2 °C using a N_2 cryostream. This thermobaric condition enabled hydrate formation in the pore space instead of ice. The target temperature was maintained for 30 h to complete the hydrate formation process [11].

2.1.1. Set-Up and Image Acquisition

Data was collected on beamline I13-2 at Diamond Light Source (DLS). Scans were performed using a polychromatic 'pink beam' at 30 keV peak energy. The detector system used was a scintillator-coupled pco.edge 5.5 camera fitted with a 4× optic magnification lens, resulting in an effective pixel size of 0.8125 μm. The X-ray projection size was 2560 × 2160 pixels (width × height).

Scans were carried out in situ at various time intervals after reaching 2 °C. The number of projections and the exposure time per projection varied amongst scans to reduce acquisition times at specific moments of the CH_4 hydrate formation process. Table 1 correlates each scan discussed in this paper with the time after the start of the 30 h sustained 2 °C period, as well as the scan specifications used.

Table 1. SXCT scan summary.

Dataset	Time at 2 °C (h)	Projections	Exposure Time per Projection (ms)
IC01	0.00	1501	200
IC02	1.53	1501	200
LC03	5.38	3001	30
LC04	10.72	3001	30
HC05	20.77	1501	30
HC06	30.02	1501	30

2.1.2. Tomographic Reconstruction and Post-Processing

Tomographic reconstruction was carried out using Savu [36–38]. Two Savu reconstruction pipelines were used: one with and one without Paganin phase enhancement [39]. These pipelines were labelled 'phase contrast' (Figure 3b) and 'absorption contrast' (Figure 3a), respectively. Both pipelines implemented filtered back-projection reconstruction [40,41] and pre-reconstruction algorithms for speckle and ring artefact suppression [37,42] and the automatic determination of the centre of rotation [43]. Further processing was carried out on the output from both reconstruction pipelines using Fiji [44,45]. This consisted in:

1. The application of a median filter of kernel size 3 to the absorption volume and the halving of the resulting greyscale values;
2. The application of an unsharp mask filter of radius 3 and weight 0.70 to the phase contrast volume;
3. The elementwise averaging of both volumes.

This procedure resulted in a single reconstructed volume with clear edge detail and phase contrast (Figure 3c).

Finally, to mitigate the halo-like or 'cupping' artefact caused by the preferential attenuation of lower-energy X-rays close to the specimen surface, known as beam hardening, as well as by truncation artefacts introduced by attenuation from sample regions outside the field of view [46,47]), each slice was convolved with two mollifier functions with an inverse shape to that of the cupping artefact. This flattened the horizontal (XY) grey value profile of each slice. A circular mask with a radius of 1100 pixels was then applied to remove voxels at the outer edges of the field of view (FOV), which were resistant to cupping correction. An example output slice is presented in Figure 3d.

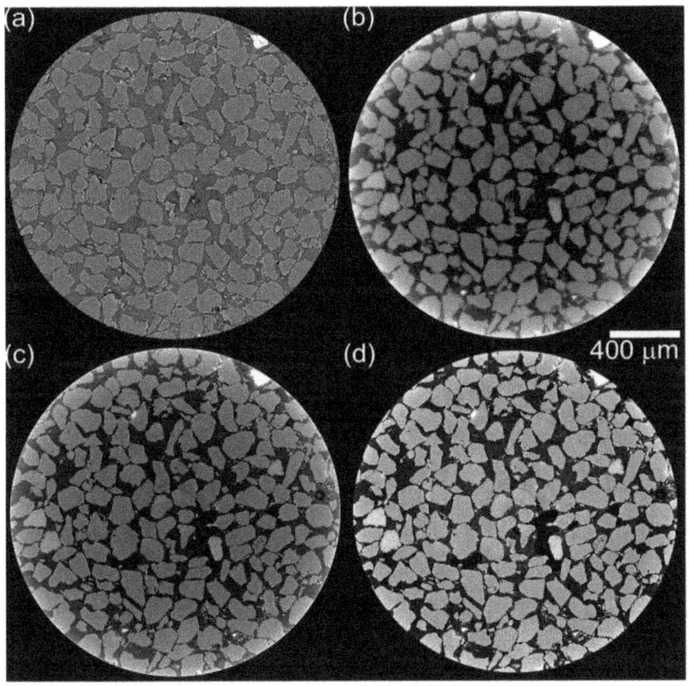

Figure 3. Slice 1050 of an intermediate-contrast SXCT dataset showing the output of the reconstruction and post-processing stages: (**a**) reconstruction through absorption contrast pipeline; (**b**) reconstruction through phase contrast pipeline; (**c**) output from filtering and volume averaging; (**d**) Cupping correction output.

As outlined in Section 1, limited greyscale contrast between the CH_4 gas and the brine-hydrate phase persisted after reconstruction and post-processing. Distinction between these two phases became increasingly difficult as the distance between the 3D image histogram peaks for the sand and non-sand phases reduced, as exemplified in Figure 1b. This distance is therefore used in this paper as an overall measure for image contrast, with regard to the ease with which the material phases could be identified and segmented. Considering this, intermediate contrast dataset IC01 89062 (Table 1) was selected initially to investigate the suitability of U-Nets to perform segmentations.

2.2. U-Net Segmentation

Three different methodologies were used to create trained U-Net models to segment the three main material phases present in the images: sand, brine-hydrates and CH_4 gas. These were:

1. A 3D hierarchical approach where two separate 3D U-Net models were trained to perform binary segmentations: On the sand phase vs. the others and the CH_4 gas phase vs. the others;
2. A 2D multi-label and multi-axis approach where a single 2D U-Net was trained to classify the three labels. The encoder section of this U-Net implementation was pre-trained on the ImageNet dataset [48], meaning that the network should only require a small amount of 'transfer' training in order to achieve acceptable results on new data;

3. RootPainter software, which uses a graphical user interface (GUI) and human intervention by interactive corrections to train a lightweight binary 2D U-Net model.

The U-Net models produced by each method were used to segment a 1554 × 1554 × 2000 voxels sized region of the 2560 × 2560 × 2000 voxels sized reconstructed and post-processed volumes, hereafter termed 'analysis region' and 'total volume'. The analysis region was inscribed within the cylindrical FOV of the total volume and omitted the black pseudo-background generated during reconstruction. Figure 4a shows the analysis region for dataset IC01. All analysis regions discussed in this paper are available in Alvarez-Borges et al. [49]. It is emphasised that the segmentation of the sand phase via U-Nets was done to assess the multi-label segmentation capacity of the algorithm. Due to its uniformly high-contrast and well-defined edges, sand could be easily segmented with any 'conventional' method, for example, thresholding.

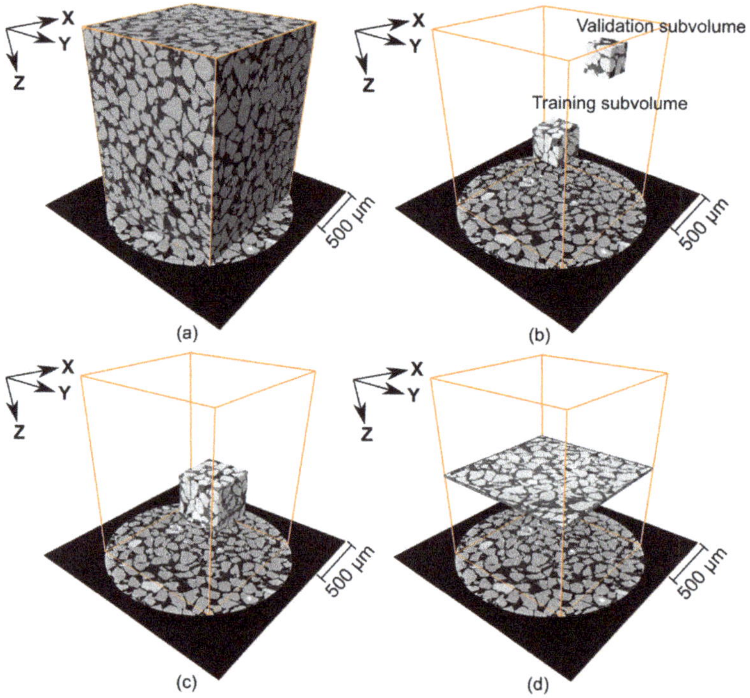

Figure 4. For dataset IC01: (**a**) analysis region used for U-Net segmentation; (**b**) location of 3D training and 3D validation subvolumes; (**c**) location of 2D learning subvolume; (**d**) quantitative analysis slices.

2.2.1. Training and Validation Data

The U-Net training procedures required both greyscale and label datasets. The latter was the 'ground truth' information used during training and validation. Label data was produced by hand-annotating the sand, CH_4 gas and brine-hydrate in the greyscale data using Avizo Lite® software. This was carried out on small subregions of the analysis volumes to reduce labelling time. The 3D hierarchical approach initially used a 384 × 384 × 384 voxels sized training subvolume and a 256 × 256 × 256 voxels sized validation subvolume, selected from two different regions of the 3D image (Figure 4b). These are hereafter referred to as 3D training and 3D validation subvolumes, respectively. RootPainter requires 2D label images (slices) of at least 572 × 572 pixels in size for both training and validation, as explained later in Section 2.2.4. Therefore, a 572 × 572 × 572 voxels sized subvolume was delimited for this purpose (Figure 4c), hereafter referred to as 2D learning subvolume. The same 2D learning subvolume was used only to train the 2D multi-label models, with the 3D validation subvolume used for validation as with the 3D methods.

The 2D and 3D training and validation subvolume coordinate origins relative to the global origin of the total volume are listed in Table 2 (note in column three that all subvolumes were 3D arrays, despite their naming and usage). The global coordinate system origin is indicated in Figure 4. All training, validation and segmented data used in this investigation are available in [49].

Table 2. U-Net training and validation subvolume details (shown in Figure 4).

Subvolume Name	Usage	Size (Voxels)	X	Y	Z	U-Net Method
3D validation	validation	256 × 256 × 256	1133	1753	50	3D hierarchical
3D training	training	384 × 384 × 384	1343	943	1158	3D hierarchical
2D learning	training & validation	572 × 572 × 572	1343	943	1158	RootPainter; 2D multilabel; 3D hierarchical *

* For additional comparison with 2D methods.

2.2.2. 3D Hierarchical Segmentation

The 3D hierarchical U-Net model was implemented in the Python library PyTorch [50] and based upon an existing implementation of a residual 3D U-Net from the literature [51,52]. This model had 35.3 million trainable parameters and five downsampling and upsampling stages. Voxel intensity values were rescaled and clipped, truncating values beyond 2.575 standard deviations of the mean to mitigate the skewing effect of outliers. The ground truth label volumes (3D training subvolume with three labels: sand, brine-hydrates and CH_4 gas) were used to create separate binary label volumes, one with sand vs. background and the other with CH_4 gas vs. background. These volumes were used as the label data for training the separate binary 3D U-Net models.

Unlike the multilabel 2D U-Net implementation described later, this model had not been pre-trained on ImageNet and was therefore likely to require a larger amount of training data to reach a high segmentation accuracy. To overcome this, the TorchIO library [53] was used to sample 128 × 128 × 128 voxels sized regions from the greyscale 3D training subvolume and generate 48 sets with random noise, flips, blurs, affine, and elastic transformations to be used as an extended training data set for each training epoch (i.e., a full training cycle). This procedure is termed 'augmentation'. These 128 × 128 × 128 voxels regions matched the input size of the U-Net.

During training, U-Net model parameter optimisation (i.e., the process of updating the model parameters on each training iteration) was carried out with a method known as AdamW [54]. The learning rate, a parameter that controls the step size of the updates made by the optimiser, was initially determined automatically and then cycled up and down every epoch to reduce the need to tune this parameter and to accelerate the training process [55]. The parameters β1, β2 and weight decay were set at 0.9, 0.999 and 0.1, respectively. Binary cross entropy (BCE), a measure of the uncertainty between two data distributions, was used as the loss function (the function minimised by the optimiser during training). Training progress was monitored using Intersection Over Union (IOU) on the validation set as the evaluation metric. If either no improvement in validation loss occurred after 40 passes of the entire training dataset (epochs) or 100 epochs were completed, the model with the lowest validation loss was saved. This was aimed at preventing overfitting. Software source code for this method is available from King and Alvarez-Borges [56].

When predicting the segmentation of the analysis region, two binary predictions were produced for each data set, one for sand vs. background and the other for CH_4 gas vs. background. These two label volumes were then combined using a label hierarchy: first, a new volume was created with all voxel labels set to brine-hydrates, then the labels corresponding to CH_4 gas were transferred from the CH_4 vs. background prediction, and lastly the labels corresponding to sand were transferred from the sand vs. background prediction.

2.2.3. 2D Multi-Label Segmentation

Training of the 2D U-Net with multiple labels was performed using the 2D learning-subvolume using two approaches. The first mimicked that of RootPainter, described later, with the network being trained on horizontal 2D (XY) slices through the image volume. The second, multi-axis approach, utilised slices taken in the XY, XZ and YZ planes (coordinate system shown in Figure 4). A 2D U-Net was used with a ResNet34 encoder [57]. This model had a total of 41.2 million trainable parameters and four downsampling and upsampling stages. This encoder was loaded with pre-trained weights from ImageNet. The model was created with Fastai [58], a Python library which has a high-level interface that utilizes PyTorch. Default Fastai image augmentations were used during training. These consisted of random image crops, zooms, rotations, flips, affine transforms and brightness and contrast adjustments. The loss function used was cross entropy (CE) and the evaluation metric used was the number of correctly labelled voxels expressed as a percentage. Training was carried out for 15 epochs.

For the single-axis implementation, the XY greyscale stack and corresponding label stack of 572 images were randomly split into training (80%) and validation (20%) sets. The input image size for the model was also set to 572×572 pixels to match that used by RootPainter. When predicting the segmentation for the analysis region, data was fed into the network in the form of 2000 XY slices of size 1554×1554 pixels.

For the multi-axis approach, the 2D greyscale learning subvolume and corresponding label data were sliced into 2D images in the XY, XZ and YZ planes, resulting in 1716 (572×3) training image and label pairs. These images were also randomly split into a training (80%) and validation (20%) set. The input image size for the model was again set to 572×572 pixels. When predicting the segmentations for the analysis region, an averaging approach for data produced from each plane was used as described by Tun et al. [59], but with a modification to take the multiple labels into account. In short, this averaging approach consisted in slicing, segmenting, and rotating the volume across the XY 4-fold symmetry plane and then splitting and hierarchically recombining the 12 resulting segmentation volumes so that two label volumes were obtained, one containing labels for sand vs. background and the other for CH_4 vs. background. These two binary label volumes were then combined into a multi-label volume as done for the data output from the 3D hierarchical method (Section 2.2.2).

Software source code for this method is available from King and Alvarez-Borges [56].

2.2.4. RootPainter Segmentation

RootPainter [60] is a client-server application originally developed to segment plant root features from photographs of soil profiles [61,62]. The client GUI is employed to annotate 2D images from a dataset, such as a tomography image stack of horizontal (XY) slices, as in the present case. The tomography slices and corresponding annotations are then read by the server and used to train the segmentation model using a U-Net variant with 1.3 million trainable parameters and four downsampling and upsampling stages. This is implemented in PyTorch and described by Smith et al. [61] and Smith et al. [62]. To execute the training routine, the software creates a validation dataset by randomly selecting one annotation image out of every five created. The accuracy of the model produced at the end of each training epoch is evaluated using the F-score parameter described by Smith et al. [62]. At the end of each training epoch, F-score values for the current and previous model are compared and the one with the highest value is saved. Training is stopped if 60 epochs are completed without F-score improvements.

RootPainter uses interactive corrections. These are created by annotating image slices overlaid with the segmentation labels produced by the best model currently available.

These corrective annotation slices are added to the training and validation datasets so that the five to one ratio is maintained.

RootPainter 0.2.5 can only predict binary segmentations ('foreground' vs. 'background'). Therefore, it was initially used to segment the CH_4 gas phase only. The 2D learning sub-volume was used for training and validation.

Sparse annotations have been shown to produce better results than dense/intensive annotations when interactively training U-Net models [61,63]. Thus, arbitrarily sparsely annotated images were produced by converting all CH_4 gas labels into foreground and enclosing them with background labels that included brine-hydrate and sand pixels, as shown in Figure 5a,b. This was done by morphologically dilating the CH_4 label of each slice in the training dataset and re-labelling the added pixels as background. The annotated slices were then copied into annotation and validation directories, maintaining the five-to-one ratio. Training was initiated after copying the first batch of five images. Further batches were added if a training epoch finished without further improvements in F-score and the model could not segment the majority of CH_4 pixels, or if the erroneously segmented pixels were patently greater than the number of correctly segmented pixels, as shown in Figure 5c. Corrective annotation was started after a training epoch had produced a model that segmented most of the CH_4 regions with a roughly equivalent number of erroneously labelled pixels, as presented in Figure 5d,e. Once a model was produced that could segment CH_4 without evident erroneously labelled pixels, the software was left to carry on training until the 60-epoch limit was reached. The resulting model was then used to segment the analysis region slice by slice.

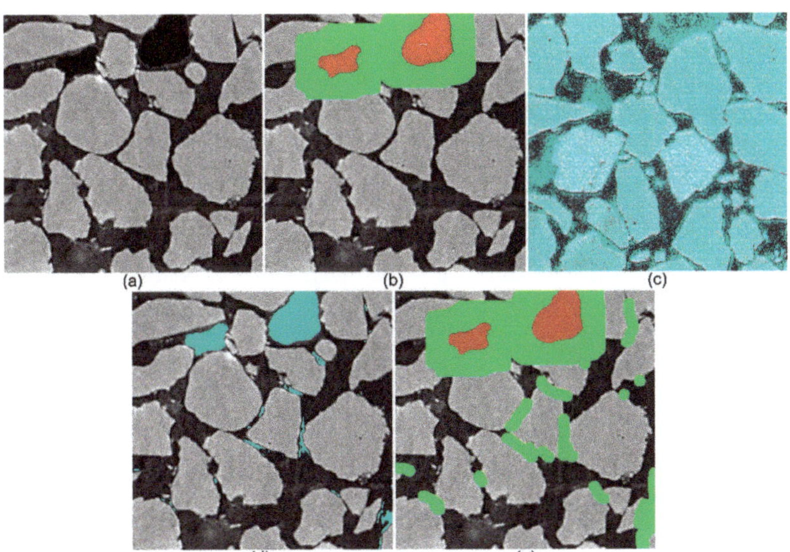

Figure 5. RootPainter usage example (on IC01 data): (**a**) XY slice from 2D learning subvolume; (**b**) slice annotations used from training and validation with CH_4 (foreground) shown in red and background shown in green; (**c**) initial segmentation output with a large number of erroneously labelled voxels; (**d**) improved segmentation with a small number of erroneously labelled voxels; (**e**) annotative correction of mislabelled voxels.

2.3. Thresholding and Watershed Segmentation

To compare the performance of the U-Net methods with conventional segmentation routines, the SXCT data was segmented using manual and automatic thresholding, and the watershed method. Images were downsampled to 8-bit as in the U-Net methods described previously. A bilateral filter was used before segmentation to improve thresholding performance and mitigate over-segmentation (filter parameters were:100-pixel spatial kernel, 50-pixel window size, and a grey-value kernel of 30 counts; implemented in Python using the open-cv library, Bradski [64]; see, e.g., Paris et al. [65] for filter description).

Manual thresholding was carried out by selecting a single threshold value for all slices by visual inspection. Automatic thresholding was performed on a slice-by-slice basis using the multi-level Otsu method [66] implemented using the scikit-image Python library [67].

Watershed segmentation was carried out in Fiji using the morphological segmentation tool in the Morpholibj library [68]. It consisted in the application of a morphological gradient with radius of 1 and the automatic determination of markers by finding local minima (with a tolerance of 8 greyscale intensity values), prior to the watershed 'inundation' phase. The output label image contained different labels for all features in the greyscale input image, including sand and brine-hydrate. Labels corresponding to regions in the 8-bit volume with mean greyscale intensity values below 50 to 70, depending on the dataset, and above 130 were classified as CH_4 gas and sand, respectively. The remaining voxels were classified as brine-hydrates.

2.4. Quantitative Analysis

The central 40 XY slices of the segmented analysis region were compared with hand-annotated counterparts created in Avizo Lite® and considered to represent 'ground truth' labels. These slices do not intersect any of the training or validation subvolumes. These ground truth volumes are available in Alvarez-Borges et al. [49]. The previously mentioned IOU metric was used to evaluate segmentation performance. IOU is defined as:

$$\mathrm{IOU} = \frac{\mathrm{TP}}{\mathrm{TP} + \mathrm{FN} + \mathrm{FP}}, \quad (1)$$

where TP refers to the number of voxels or pixels correctly predicted to correspond to the label of interest ('true positive'), and FP and FN are the number of voxels or pixels incorrectly predicted to be part of the label of interest ('false positive') and voxels/pixels incorrectly predicted to belong to any of the other material phases ('false negative'), in each case. A comparable analysis of U-Net accuracy has been done by, e.g., Karabağ et al. [69] and Phan et al. [32].

IOU returns a value between 0 and 1, where the latter corresponds to the scenario were the segmentation matches the validation image pixel by pixel (or voxel by voxel). In the following sections, quantitative analyses were carried out on a slice-by-slice basis (i.e., using pixel counts as input).

3. Results and Discussion

3.1. Segmentation Performance Comparison

Figure 6 compares the original and segmented central slice for dataset IC01, produced using the three U-Net methods (Section 2.2) and three standard methods (Section 2.3). Training and validation in both the 2D multi-label approach and RootPainter was carried out using XY slices only (i.e., single plane). Figure 7a presents accuracy metrics for the segmentation of CH_4 gas in the central 40 XY slices this dataset. It may be noted that RootPainter delivered slightly higher metrics than the other two U-Net methods, but this difference in performance cannot be readily identified in Figure 6. Figures 6 and 7a also show that, for this dataset, watershed and manual thresholding methods return lower accuracy results than the U-Net approaches, and that Otsu-thresholding performed poorly. In fact, the Otsu approach consistently segmented the brine-hydrate and CH_4 gas as a single label, as evident in Figure 6e. This is chiefly due to the absence of well-defined inter-class variance extrema between these materials and the small relative size of CH_4 bubbles [70,71]. In later comparisons, results from the Otsu method are omitted for this reason.

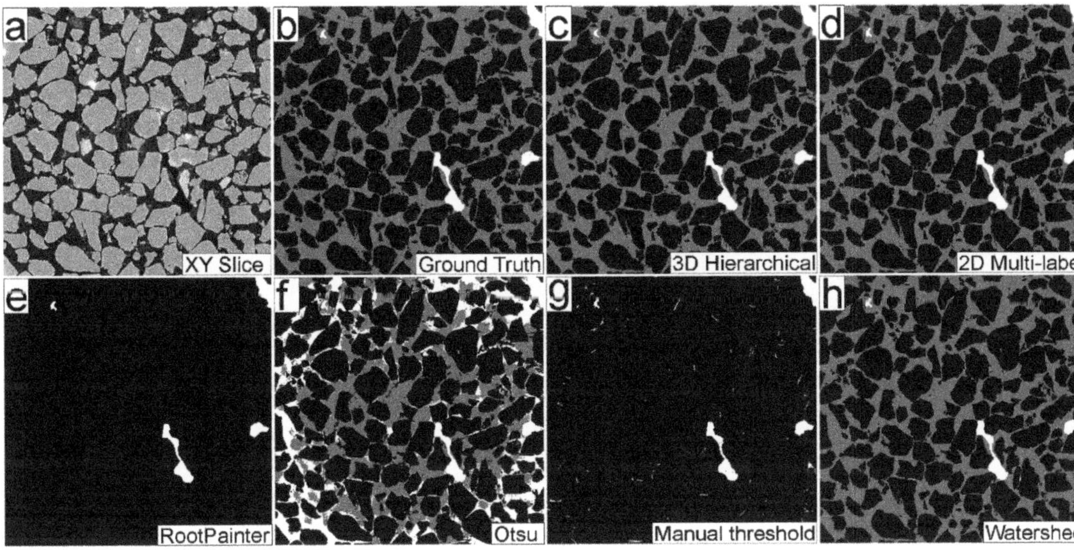

Figure 6. (**a**) Original XY central slice of dataset IC01; (**b**) hand-annotated ground truth labelled slice; (**c**) segmented slice using the 3D hierarchical method; (**d**) segmented slice using the 2D multilable singe-axis approach; (**e**) RootPainter segmentation of the CH_4 gas phase; (**f**) Otsu auto-threshold output; (**g**) manual thresholding output; (**h**) Watershed segmentation. CH_4 gas shown in white, brine in grey and sand in black.

Since each model had been trained to convergence with their respective data, the slightly lower performance metrics observed in Figure 7a for the 3D hierarchical output, compared to that of RootPainter, may be attributed to the smaller training subvolume used. To present a more balanced comparison, a further 3D hierarchical model was trained on a subvolume of the same size as the one used for both 2D methods, i.e., the larger 2D learning subvolume. This comparison is presented in Figure 7b, where it is evident that RootPainter still outperformed the 3D hierarchical approach, though the difference between methods reduced. While an even larger training subset may deliver more substantial improvement, the preparation of such data would require much greater operator input, which is contrary to the aim of this study.

Figure 7a,b show that pre-training on the ImageNet database for the 2D multi-label method did not result in a significant segmentation performance advantage over the 3D hierarchical method. A similar outcome on the effect of transfer learning has been reported by He et al. [72]. They remarked that, ultimately, pre-training primes the U-Net for feature identification, which leads to fewer training iterations rather than greater segmentation accuracy. Such appears to be the present case, as the 2D multi-label approach produced similar results to the 3D hierarchical method with up to six times fewer training epochs, as shown in Tables 3 and 4.

A disadvantage of the use of 2D U-Net segmentation methods that operate solely with XY slices, such as RootPainter and the single-axis 2D multi-label method, is that horizontal stripe artefacts may appear in the vertical (YZ or XY) slices of the segmented volume. This occurs because training and segmentation does not account for feature continuity between slices. Such artefacts are absent in the output of the 3D hierarchical implementation, which is reflected in the "smoothness" of the line showing the per-slice metrics for this approach in Figure 7. These artefacts can be mitigated by predicting segmentation of data slices taken along different axes and subsequently recombining them into a single volume, as done for the multi-axis 2D method, described in Section 2.2.3. This also improves the algorithm segmentation performance metrics, as shown in Figure 7b, but at the expense of greater computation times, as presented in Figure 8.

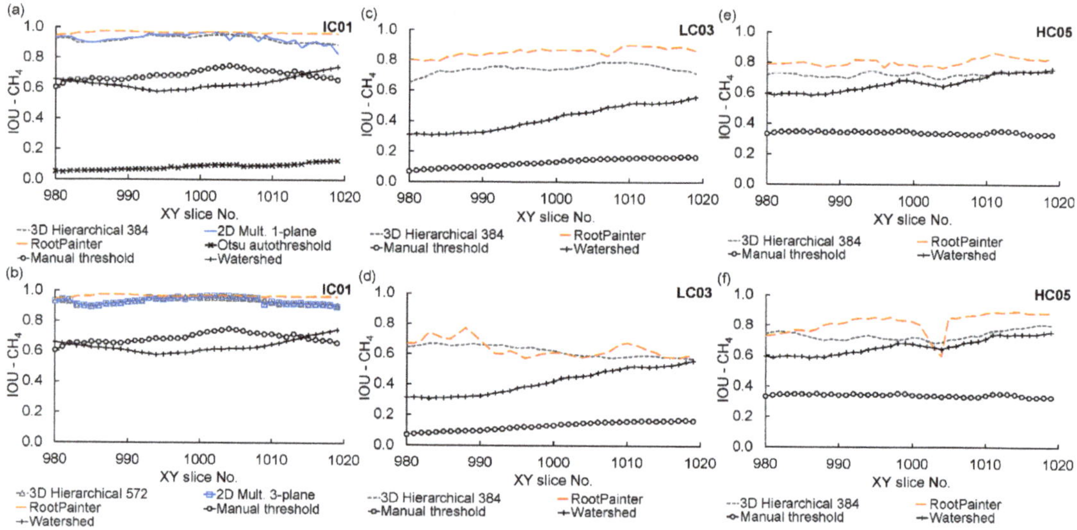

Figure 7. Performance metrics for the segmentation of CH_4 gas on the central 40 XY slices of: (**a**) IC01 using the 3D hierarchical, the single-axis 2D multilabel and RootPainter U-Nets; (**b**) IC01 using the 3D hierarchical (with the 2D learning subvolume), the multi-plane 2D multilabel and RootPainter U-Nets; (**c**) LC03 using the 3D hierarchical and Rootpainter U-Nets; (**d**) LC03 using the 3D hierarchical and RootPainter U-Nets trained on subvolumes from IC01; (**e**) HC05 using the 3D hierarchical and RootPainter U-Nets trained on subvolumes from IC01; (**f**) HC05 using the 3D hierarchical and RootPainter U-Nets trained on subvolumes from LC03. Watershed and thresholding results shown for reference.

Table 3. Binary 3D hierarchical U-Net training metrics.

Training Data Source	Labels	Training Epochs	Final Training Loss (BCE)	Final Validation Loss (BCE)	Final Validation Metric (Mean IOU)
IC01–3D training subvolume	CH_4 vs. background Sand vs. background	94 83	0.0313 0.0317	0.0237 0.0332	0.935 0.977
IC02–2D learning subvolume	CH_4 vs. background Sand vs. background	84 85	0.0055 0.0426	0.0307 0.0290	0.918 0.980
LC03–3D training subvolume	CH_4 vs. background Sand vs. background	82 69	0.0178 0.0305	0.0371 0.0471	0.759 0.957

Table 4. Single- and multi-axis 2D multilabel U-Net training metrics.

Training Data Source	Labels	Training Epochs	Final Training Loss (CE)	Final Validation Loss (CE)	Final Validation Metric (%) *
IC02–2D learning subvolume	Single-axis (pre-trained) Multi-axis (pre-trained)	15 15	0.0200 0.0190	0.0140 0.0120	99.48 99.59

* %–Percentage of correctly labelled voxels.

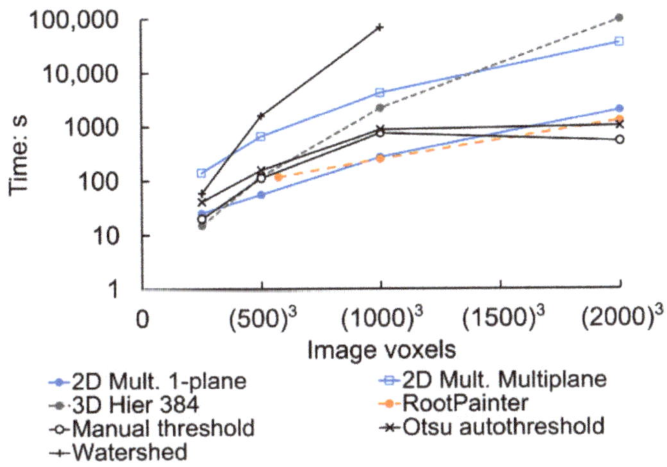

Figure 8. Segmentation time required using an Nvidia Tesla V100® GPU for the U-Net methods and HPCC of 40×Intel Gold 6242R® CPU @ 3.10 GHz for the standard methods. Benchmarking data were extracted from the reconstructed and post-processed scan IC01 and are available from Alvarez-Borges et al. [49]. Excludes time spent on the preparation of training datasets.

3.2. U-Net Performance on Data with Different Greyscale Contrast

The trained U-Net models created with all three methods described above were able to predict high-quality segmentations for the intermediate contrast dataset IC01 when trained on subsections of the same dataset. To examine if similar results could be obtained on datasets exhibiting lower greyscale contrast, both 3D hierarchical and RootPainter U-Nets were used to segment 'low' contrast dataset LC03 (Table 1, Figure 1b), using 3D training and 2D learning subvolumes of the same data for training, respectively. Figure 7c presents the performance metrics resulting from this approach, as well as those of watershed and manual thresholding methods applied to the same volume. It may be noted that both U-Net methods return lower metrics than those used on IC01. The IOU computations show that, on average, 74% and 85% of the voxels predicted to be CH_4 gas were true positives in the 3D hierarchical and RootPainter results, respectively. In comparison, these average values were 92 and 94% for IC01.

Figure 9a shows that, for both U-Net methods, the lower performance metrics of the segmentation for LC03 are driven by false positives. However, false positives are over twice as numerous than false negatives in the results for the 3D hierarchical approach, whereas they only surpass false negatives by about 30% in the RootPainter segmentation. For both methods, most false positives correspond to ground truth brine-hydrate voxels incorrectly labelled as CH_4 gas, as depicted in Figure 9b. This indicates that the reduced grey value differentiation (i.e., contrast) between CH_4 gas and brine-hydrate phases restricted U-Net segmentation accuracy, as anticipated. Despite this, the U-Net methods significantly outperform the standard approaches, as shown in Figure 7c. In fact, performance numbers reveal that watershed and manual thresholding cannot deliver a reliable quantification of the material phases of this low contrast dataset.

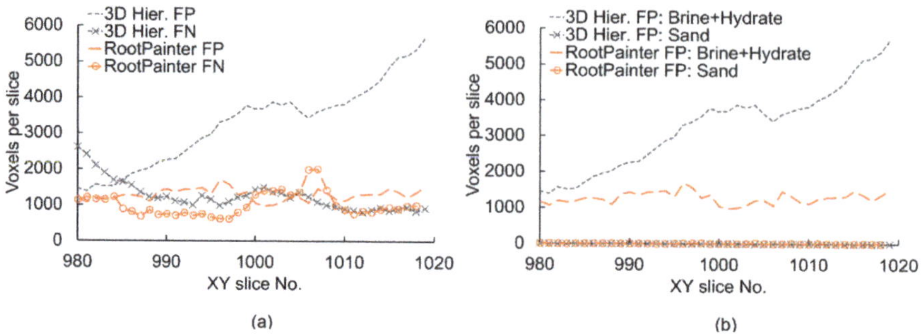

Figure 9. (**a**) False positive (FP) and false negative (FN) CH_4 gas voxels and (**b**) FP voxel labels per slice for the central 40 slices of dataset LC03 segmented using RootPainter and the 3D hierarchical method (3D Hier).

3.3. U-Net Segmentation Model Generalisation across Datasets (Model Portability)

To examine U-Net model portability, 'low' and 'high' contrast datasets LC03 and HC05 (Table 1, Figure 1b) were segmented using the models produced from training on 'intermediate' contrast dataset IC01. Figure 7d,e presents the performance metrics of the resulting segmentations. It can be observed that segmentation accuracy is lowest in the case where the U-Net models trained on mid-contrast dataset IC02 were applied to the low-contrast dataset LC03. IOU values from this process are comparable to those obtained from the thresholding and watershed methods applied to mid-contrast dataset IC01, and thus, quantification from these segmentations may be unreliable. The U-Net model trained on IC01 produced higher accuracy segmentations of high-contrast dataset HC05, comparable to those for the segmentation of low-contrast dataset LC03 using models trained 'natively' on LC03 subvolumes. Yet, it is evident that U-Net models trained on subvolumes of IC01 perform best when applied to the same 'native' IC01 dataset, as shown by Figure 7a,e.

As segmentation performance appeared to be higher when U-Net models trained on lower contrast data were used to segment higher contrast data, models trained on low-contrast LC03 images were used to segment high-contrast dataset HC05. Performance metrics are presented in Figure 7f. This Figure shows an overall improvement in performance metrics compared with segmentations produced with the U-Net models trained on IC01 (Figure 7e). However, an instance of localised poor performance for RootPainter can be observed in the profiles of Figure 7f, which resulted from a cluster of FP pixels on a single slice. This emphasises the limitations of the slice-by-slice (2D) segmentation described in Section 3.1, and denotes a broadly similar pattern of FP-driven model inaccuracy as for the results discussed previously in Section 3.2 (Figure 9).

3.4. Applications and Implications

The segmentation of XCT or SXCT images of soil and rock samples is often carried out to determine parameters such as porosity or liquid/gas saturation, as discussed in Section 1. The varying performances of the U-Net methods used in the present investigation result in differences in the parameters calculated from the segmented images. This is exemplified in Figure 10, which compares porosity and CH_4 gas saturation (by volume) ratios derived on a slice-by-slice basis from the segmented volumes produced with the 3D hierarchical approach (3D training sub-volume) and RootPainter, which were the procedures that seemed to provide the best results with the least user time. Calculation details are provided in Appendix A.

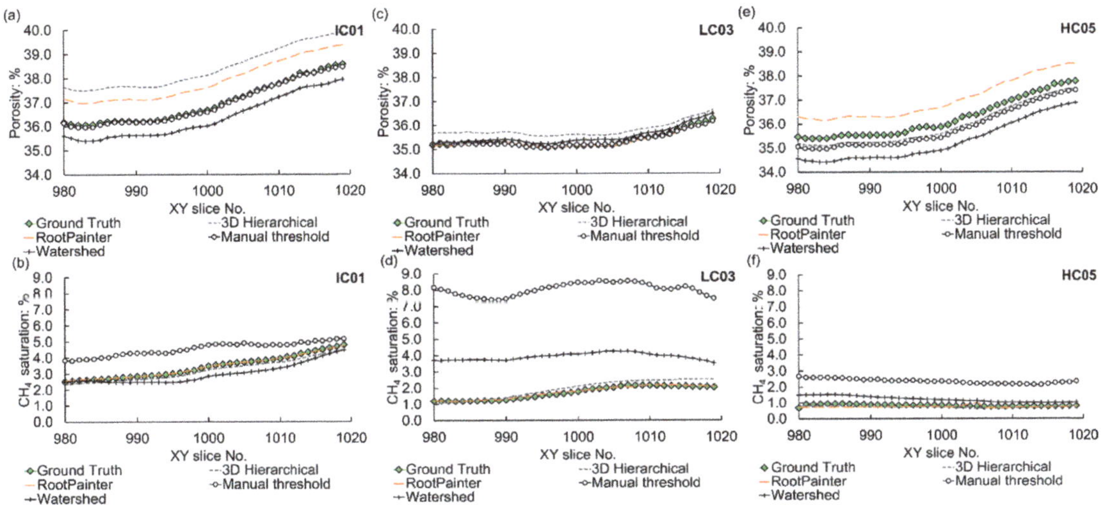

Figure 10. Porosity and CH_4 gas saturation profiles for the central 40 XY slices of data sets IC01 (**a**,**b**), LC03 (**c**,**d**) and HC05 (**e**,**f**) derived using image segmentations obtained from 3D hierarchical and RootPainter U-Net models trained on sub-volumes of IC01 (**a**,**b**) and LC03 (**c**–**f**). Parameters derived from manual thresholding and watershed segmentation methods also shown.

Results presented in Figure 10 correspond to two application scenarios, that is:

1. U-Nets trained on sub-volumes of the dataset of interest are then used to segment the entire dataset, shown in Figure 10a,d. As discussed in Section 3.2, differences in greyscale contrast affect the performance of the resulting segmentation. A training sub-volume needs to be created for each scan;
2. U-Nets trained on sub-volumes of a low-greyscale contrast dataset are then used to segment other 'unknown' datasets of higher greyscale contrast (model portability). This is presented in Figure 10e,f, corresponding to parameters derived for high-contrast dataset HC05 using segmentations produced from U-Nets trained on sub-volumes of low-contrast dataset LC03. Thus, only one training sub-volume is needed to segment multiple scans.

Porosity and CH_4 saturation calculations derived from manual thresholding and watershed methods are also included in Figure 10. This Figure suggests that, while U-Net models trained on a sub-volume of the same data delivered high segmentation performance metrics for the CH_4 gas phase, the derived parameters deviated from ground truth values to some extent, this being more acute for porosity inferences. In fact, in most cases watershed or thresholding methods delivered more accurate porosity profiles. A comparison between the mean absolute error (MAE) for the porosity and CH_4 saturation calculations along with the mean IOU values for the combined CH_4 gas and sand labels from the three volumes used to generate Figure 10 is shown in Figure 11. This Figure reveals that, while there is a general trend of lower MAE for derived material parameters with higher segmentation accuracy, the correlation exhibits some scatter. Considering that both CH_4 gas saturation and porosity are in part derived using the number of sand voxels and that these are significantly more numerous than pore voxels (CH_4 gas and brine-hydrates), it may be proposed that errors in porosity/CH_4-saturation estimation originate from inaccuracies in the segmentation of the sand phase. This is evidenced in Figure 12 for dataset IC01, which presents (a) IOU metrics for the segmentation of the sand phase and (b) the number of FP and FN voxels. Figure 12a reveals that the inaccuracies in the segmentation of the sand phase are relatively small in terms of metrics, which are in fact higher than those of the CH_4 gas phase presented in Figure 7a. However, Figure 12b shows that the number of FP and

FN voxels is large compared to the size of the CH_4 gas and brine-hydrate phases, which amount to roughly 3.0×10^4 and 8.75×10^5 voxels per slice, respectively. This, in turn, affects parameters calculated from voxel counts. This denotes that the estimation of soil parameters based on ratios between material phases from segmented images is particularly sensitive to the relative size of said phases. It should be noted, however, that the maximum absolute errors presented in Figure 11 for U-Net-derived parameters (1.40% and 0.26% for porosity and CH_4 gas saturation, respectively) are smaller than those commonly reported for laboratory methods [73–75].

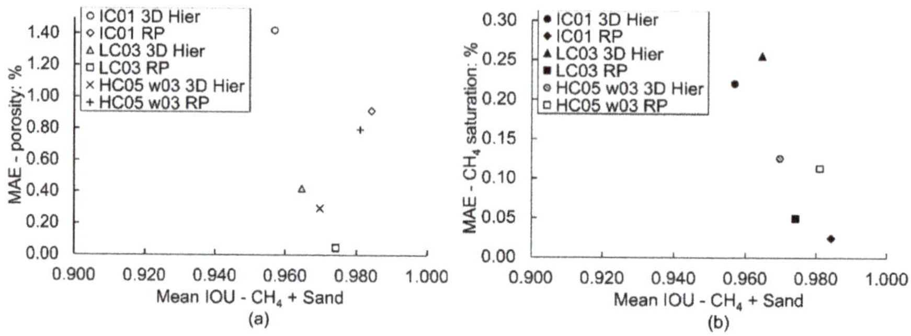

Figure 11. Comparison of mean absolute errors for (**a**) porosity and (**b**) CH_4 gas saturation estimations with mean IOU metrics for the segmentations used. W03 denotes the use of a U-Net model trained on a sub-volume of low-contrast dataset LC03; RP refers to RootPainter.

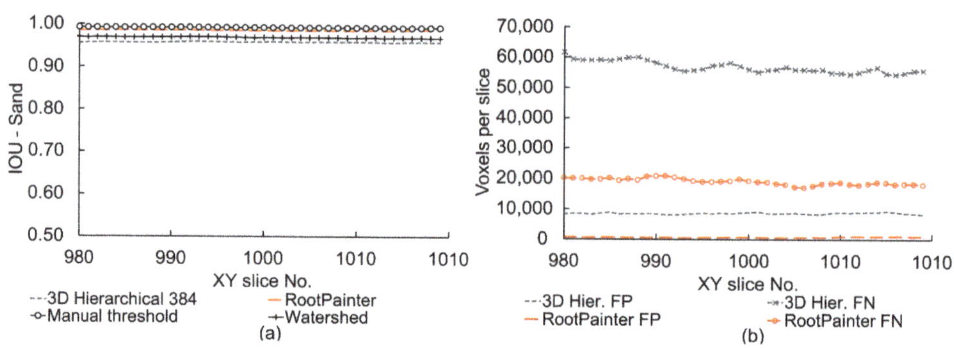

Figure 12. (**a**) IOU metrics for the segmentation of sand in dataset IC01 using the 3D hierarchical method (3D training/validation sub-volume) and RootPainter, and (**b**) associated false positive (FP) and false negative (FN) sand voxels per slice of the central 40 XY slices. Metrics for the manual thresholding and watershed methods included in (**a**) for reference.

Figure 12a also shows that thresholding and watershed methods are very effective at segmenting abundant, high-contrast, well-defined features like sand, as stated in Section 2.2. Indeed, the use of U-Nets may not be necessary or recommended if only such segmentations are required, as mentioned in Section 2.1.2. However, mainstream methods return unsatisfactory CH_4 gas saturation measurements, particularly for the low-contrast 89069 volume, as shown in Figure 10. This is due to their inability to detect scarce, low-contrast features like CH_4 bubbles, as demonstrated in Section 3.1.

A further application for U-Net segmentations of XCT/SXCT images of soil and rock is 3D data visualisation, which can then be used to investigate, for instance, CH_4 gas distribution within the pore matrix. Such application can greatly benefit from model portability. To exemplify this, Figure 13 compares 3D views of the CH_4 gas phase produced by segmenting datasets obtained at different stages of hydrate formation (Table 1) using the RootPainter model trained on the low-contrast LC03 sub-volume. Despite the presence of a modest number of segmentation errors in the form of small islands on some of the images (Figure 13b–d), the U-Net model produces sensible 3D representations of the data, and changes in CH_4 gas distribution as it is consumed for hydrate formation can be clearly distinguished. In a further example, a 2D multi-label U-Net, trained using the single-axis approach on the 2D learning subvolume from scan IC01, has been used to segment a higher-contrast SXCT scan from a similar experiment carried out at the Swiss Light Source (SLS) originally reported by Sahoo et al. [8]. The post-processing steps described in Section 2.1.2, except cupping correction, were applied to the reconstructed data and a $1554 \times 1554 \times 2000$ voxel region was extracted from the centre of the 3D image (data is available from Alvarez-Borges et al. [49]). Results are shown in Figure 14, where it is seen that the model delivers qualitatively accurate 3D views of the distribution of all three material phases, without any additional training, corrections or user input.

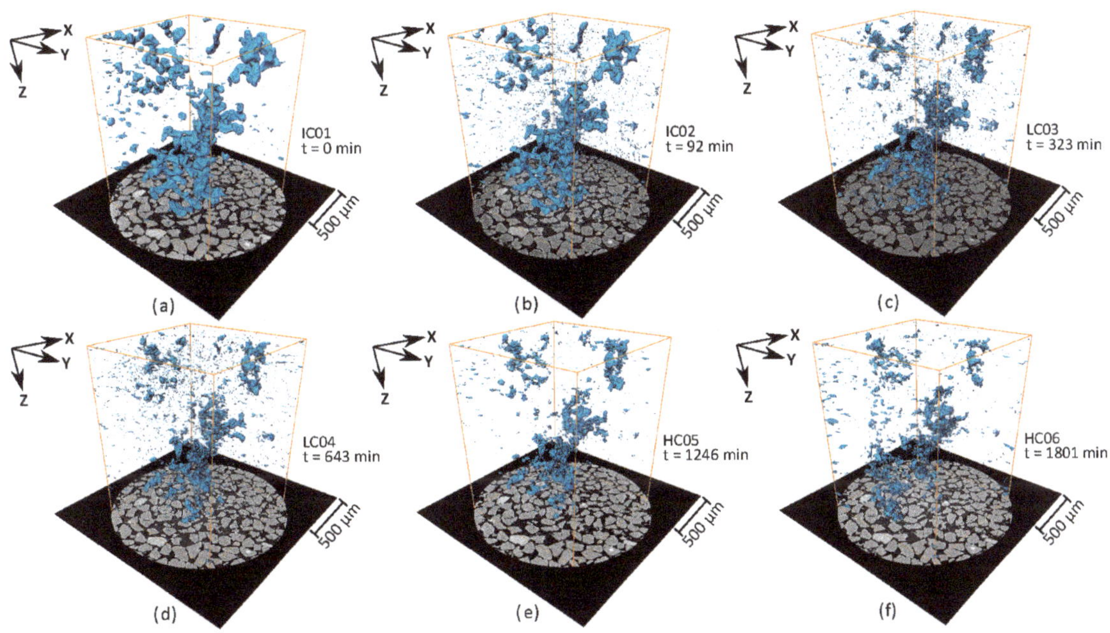

Figure 13. XCT-derived 3D views of CH_4 gas depletion during hydrate formation, with t denoting cooling time in minutes after reaching 2 °C: (**a**) t = 0, (**b**) t = 92, (**c**) t = 323, (**d**) t = 643, (**e**) t = 1246, (**f**) t = 1801. Segmentation carried out using a RootPainter U-Net model trained on the low-contrast LC03 data.

Both examples demonstrate the capability of U-Net models to segment multiple SXCT images of CH_4-bearing soil, despite being obtained using different instruments and set-ups. The U-Net models used only a single $(572)^3$ voxel sub-volume for training and did not require any additional training, corrections or user input to segment new images. A key implication is that training of a single U-Net model on a low greyscale contrast dataset could be used to deliver insight on variations in sediment morphology in other datasets. This has valuable applications. For example, segmentations are often required during a short period of time with limited operator input, such as during data acquisition at a synchrotron or other X-ray facility. The availability of pre-trained U-Net models would allow segmentations and sediment morphology/microstructure information to be produced within a short time after acquisition and reconstruction. Pre-trained models could also be used to segment numerous and/or large data sets over shorter timespans with reduced user effort and bias. However, accuracy will remain lower that what could be obtained with a model trained 'natively' on a subset of the target SXCT volume.

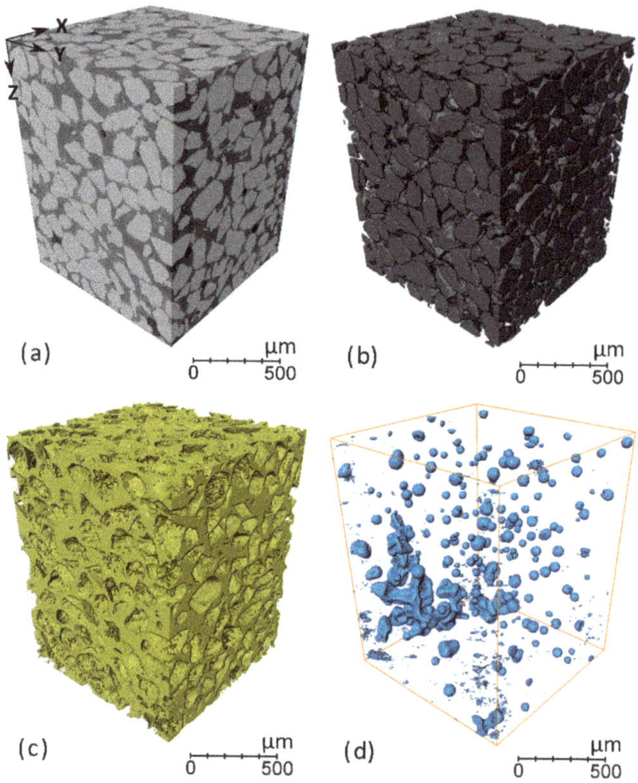

Figure 14. U-Net segmentation of an independent data set from Sahoo et al. (2018a) acquired at SLS, using a 2D multi-label single-axis U-Net model trained on the 2D learning subvolume for IC01: (**a**) reconstructed SLS volume; (**b**) sand; (**c**) brine-hydrate; (**d**) CH_4 gas.

Future work could encompass, for instance, an analysis of the effect of U-Net segmentation accuracy and training strategy on 'second order' metrics such as particle, pore and bubble morphometry, as well as the potential for improving a model's ability to generalise and accurately segment new data by including training data from several SXCT volumes with varying contrast characteristics.

4. Conclusions

The application of U-Nets to segment SXCT images of CH_4-bearing sand has been investigated. The general aim was to determine if these convolutional deep learning networks, trained on a small set of images ($\leq(572)^3$ voxels), were capable of accurately segmenting large SXCT datasets ($2000 \times (1554)^2$ voxels) of different greyscale contrast, with focus on the CH_4 gas phase. Training images were obtained from a hand-annotated subset of the reconstructed SXCT data. Three U-Net deployment methods were used: 3D hierarchical, 2D multi-label and the RootPainter application. Quantitative comparisons amongst U-Net segmentation outputs, along with mainstream thresholding and watershed methods, were carried out using the IOU metric. Major outcomes of this investigation are presented below.

1. For a given SXCT data set, the three U-Net deployment methodologies produced models capable of delivering segmented images of the CH_4 gas phase with average IOU metrics of at least 0.74 and up to 0.93. This demonstrated that the U-Net methods used were capable of accurately identifying the CH_4 gas phase using a small number of training images. RootPainter delivered marginally higher IOU metrics than the other methods but suffered from minor horizontal stripping artefacts and required more human intervention and proportionally higher computing time;
2. Greyscale contrast between material phases in the different SRXCT datasets was a significant factor affecting U-Net segmentation accuracy. The lowest segmentation performance metrics corresponded to SRXCT datasets exhibiting the lowest greyscale contrast, while greater segmentation accuracy resulted from the use of higher contrast data;
3. All U-Net segmentations of CH_4 gas outperformed thresholding and watershed methods. However, mainstream methods proved to be more accurate at segmenting abundant, well-defined, and high-contrast features, like sand. U-Net methods are, thus, not recommended for this task;
4. The ability of a U-Net model trained on a subset of one dataset to generalise and produce an accurate segmentation of a different dataset, was explored. It was found that models trained on lower-contrast images were able to produce accurate segmentations of higher-contrast data without additional training. In comparison, U-Net models trained on higher-contrast images were found to deliver poor results when used to segment lower-contrast data. 'Portability' was further demonstrated by accurately segmenting independent data from a different synchrotron facility without additional training. This suggests that targeted training on small amounts of 'ground truth' data can produce U-Net segmentation models that can be used for rapid segmentation of a large number of different datasets with additional user input or training. However, segmentation accuracy will be lower than that of a model 'natively' trained on subsets of the target dataset;
5. The effect of segmentation accuracy on image-derived material parameters was investigated by calculating porosity and CH_4 gas saturation profiles using U-Net segmentations. A general trend of lower mean absolute error of the derived parameter with greater segmentation accuracy was found, but the correlation exhibited some scatter. Considering that porosity, fluid saturation and other parameters are ratios between material phases, it was proposed that errors in derived parameters are not only linked to segmentation accuracy metrics but to the number of false positive and negative voxel labels of the largest phase relative to the other phases.

Author Contributions: Conceptualisation, O.N.F.K., F.J.A.-B. and S.I.A.; methodology, O.N.F.K., F.J.A.-B., S.I.A. and B.N.M.; software, O.N.F.K.; validation, O.N.F.K. and F.J.A.-B.; formal analysis, F.J.A.-B.; investigation, F.J.A.-B., O.N.F.K. and B.N.M.; resources, T.C., M.B. and S.I.A.; data curation, F.J.A.-B. and O.N.F.K.; writing—original draft preparation, F.J.A.-B. and O.N.F.K.; writing—review and editing, O.N.F.K., F.J.A.-B., T.C., S.I.A., B.N.M. and M.B.; visualisation, F.J.A.-B.; supervision, T.C., M.B. and S.I.A.; project administration, T.C., M.B. and S.I.A.; funding acquisition, B.N.M. All authors have read and agreed to the published version of the manuscript.

Funding: This research was funded in part by the Natural Environment Research Council (UK), grant number NE/K00008X/1, awarded to B.N.M.

Institutional Review Board Statement: Not applicable.

Informed Consent Statement: Not applicable.

Data Availability Statement: Data presented in this study is openly available (under an Apache V2 License) at https://doi.org/10.5281/zenodo.4580278 (accessed on 15 September 2022), reference number [49]. Source code for the U-Net segmentation methods is available under an Apache V2 License at https://github.com/DiamondLightSource/gas-hydrate-segmentation-unets (accessed on 14 April 2022) for the 2D Multilabel and 3D Hierarchical methods, reference number [56], and under a GNU General Public License v3.0 at https://github.com/Abe404/root_painter (accessed 15 March 2022) for RootPainter, reference number [60].

Acknowledgments: The authors gratefully acknowledge Diamond Light Source for the provision of beamtime (proposal MT16205-1) and are thankful for the support provided by beamline I13 staff.

Conflicts of Interest: The authors declare no conflict of interest. The funders had no role in the design of the study; in the collection, analyses, or interpretation of data; in the writing of the manuscript; or in the decision to publish the results.

Appendix A

Porosity was calculated as:

$$\text{Porosity (\%)} = \frac{\text{volume of pores}}{\text{total volume}} \times 100, \tag{A1}$$

and CH_4 gas saturation was determined as:

$$CH_4 \text{ saturation} = \frac{\text{volume of } CH_4}{\text{volume of pores}} \times 100, \tag{A2}$$

where the volume of CH_4 gas amounts to the total number of CH_4 gas voxels, the volume of pores is the sum of CH_4 gas and brine-hydrate voxels, and the total volume is the total number of voxels in the image multiplied by the voxel volume ($0.8125 \times 0.8125 \times 0.8125$ µm). For the RootPainter method, the sand phase has been segmented using the same approach used for CH_4 described in Section 2.2.4, but using sand labels and only one quadrant of each annotation slice to produce sparsely annotated training and validation images.

References

1. Dean, J.F.; Middelburg, J.J.; Röckmann, T.; Aerts, R.; Blauw, L.G.; Egger, M.; Jetten, M.S.M.; de Jong, A.E.E.; Meisel, O.H.; Rasigraf, O.; et al. Methane Feedbacks to the Global Climate System in a Warmer World. *Rev. Geophys.* **2018**, *56*, 207–250. [CrossRef]
2. IPCC. *Climate Change 2013: The Physical Science Basis. Contribution of Working Group I to the Fifth Assessment Report of the Intergovernmental Panel on Climate Change*; Press, C.U., Ed.; Intergovernmental Panel on Climate Change: Cambridge, UK; New York, NY, USA, 2013.
3. Kvenvolden, K.A. Gas hydrates—Geological perspective and global change. *Rev. Geophys.* **1993**, *31*, 173–187. [CrossRef]
4. James, R.H.; Bousquet, P.; Bussmann, I.; Haeckel, M.; Kipfer, R.; Leifer, I.; Niemann, H.; Ostrovsky, I.; Piskozub, J.; Rehder, G.; et al. Effects of climate change on methane emissions from seafloor sediments in the Arctic Ocean: A review. *Limnol. Oceanogr.* **2016**, *61*, S283–S299. [CrossRef]
5. Ruppel, C.D.; Kessler, J.D. The interaction of climate change and methane hydrates. *Rev. Geophys.* **2017**, *55*, 126–168. [CrossRef]
6. Sahoo, S.K.; Marín-Moreno, H.; North, L.J.; Falcon-Suarez, I.; Madhusudhan, B.N.; Best, A.I.; Minshull, T.A. Presence and Consequences of Coexisting Methane Gas With Hydrate Under Two Phase Water-Hydrate Stability Conditions. *J. Geophys. Res. Solid Earth* **2018**, *123*, 3377–3390. [CrossRef]
7. Yokohama, T.; Nakayama, E.; Kuwano, S.; Saito, H. Relationship between seismic wave velocities, eletric resistivities and saturation ratio of methane hydrate using core samples in laboratory experiments. In Proceedings of the 7th International Conference on Gas Hydrates, Edinburgh, UK, 17–21 July 2011; pp. 1830–1833.
8. Sahoo, S.K.; Madhusudhan, B.N.; Marín-Moreno, H.; North, L.J.; Ahmed, S.; Falcon-Suarez, I.H.; Minshull, T.A.; Best, A.I. Laboratory Insights into the Effect of Sediment-Hosted Methane Hydrate Morphology on Elastic Wave Velocity from Time-Lapse 4-D Synchrotron X-Ray Computed Tomography. *Geochem. Geophys. Geosyst.* **2018**, *19*, 4502–4521. [CrossRef]
9. Moridis, G.; Collett, T.S.; Pooladi-Darvish, M.; Hancock, S.H.; Santamarina, C.; Boswell, R.; Kneafsey, T.J.; Rutqvist, J.; Kowalsky, M.B.; Reagan, M.T.; et al. Challenges, Uncertainties, and Issues Facing Gas Production from Gas-Hydrate Deposits. *SPE Reserv. Eval. Eng.* **2011**, *14*, 76–112. [CrossRef]
10. Saunois, M.; Stavert, A.R.; Poulter, B.; Bousquet, P.; Canadell, J.G.; Jackson, R.B.; Raymond, P.A.; Dlugokencky, E.J.; Houweling, S.; Patra, P.K.; et al. The Global Methane Budget 2000–2017. *Earth Syst. Sci. Data* **2020**, *12*, 1561–1623. [CrossRef]
11. Madhusudhan, B.N.; Clayton, C.R.I.; Priest, J.A. The Effects of Hydrate on the Strength and Stiffness of Some Sands. *J. Geophys. Res. Solid Earth* **2019**, *124*, 65–75. [CrossRef]
12. Song, Y.; Luo, T.; Madhusudhan, B.N.; Sun, X.; Liu, Y.; Kong, X.; Li, Y. Strength behaviors of CH4 hydrate-bearing silty sediments during thermal decomposition. *J. Nat. Gas Sci. Eng.* **2019**, *72*, 103031. [CrossRef]

13. Maslin, M.; Owen, M.; Betts, R.; Day, S.; Jones, T.D.; Ridgwell, A. Gas hydrates: Past and future geohazard? *Philos. Trans. R. Soc. A* **2010**, *368*, 2369–2393. [CrossRef] [PubMed]
14. Mienert, J. Methane Hydrate and Submarine Slides. In *Encyclopedia of Ocean Sciences*, 2nd ed.; Steele, J.H., Ed.; Academic Press: Oxford, UK, 2009; pp. 790–798.
15. Vanneste, M.; Sultan, N.; Garziglia, S.; Forsberg, C.F.; L'Heureux, J.-S. Seafloor instabilities and sediment deformation processes: The need for integrated, multi-disciplinary investigations. *Mar. Geol.* **2014**, *352*, 183–214. [CrossRef]
16. Holland, M.; Schultheiss, P. Comparison of methane mass balance and X-ray computed tomographic methods for calculation of gas hydrate content of pressure cores. *Mar. Pet. Geol.* **2014**, *58*, 168–177. [CrossRef]
17. Kerkar, P.B.; Horvat, K.; Jones, K.W.; Mahajan, D. Imaging methane hydrates growth dynamics in porous media using synchrotron X-ray computed microtomography. *Geochem. Geophys. Geosyst.* **2014**, *15*, 4759–4768. [CrossRef]
18. Lei, L.; Seol, Y.; Jarvis, K. Pore-Scale Visualization of Methane Hydrate-Bearing Sediments With Micro-CT. *Geophys. Res. Lett.* **2018**, *45*, 5417–5426. [CrossRef]
19. Fonseca, J.; O'Sullivan, C.; Coop, M.R. Image Segmentation Techniques for Granular Materials. In *Powders and Grains: Proceedings of the 6th International Conference on Micromechanics of Granular Media*; American Institute of Physics: Melville, NY, USA, 2009; pp. 223–226.
20. Iassonov, P.; Gebrenegus, T.; Tuller, M. Segmentation of X-ray computed tomography images of porous materials: A crucial step for characterization and quantitative analysis of pore structures. *Water Resour. Res.* **2009**, *45*. [CrossRef]
21. Rogowska, J. 5—Overview and Fundamentals of Medical Image Segmentation. In *Handbook of Medical Imaging*; Bankman, I.N., Ed.; Academic Press: San Diego, CA, USA, 2000; pp. 69–85.
22. Zhang, X.; Jia, F.; Luo, S.; Liu, G.; Hu, Q. A marker-based watershed method for X-ray image segmentation. *Comput. Methods Programs Biomed.* **2014**, *113*, 894–903. [CrossRef]
23. Baveye, P.C.; Laba, M.; Otten, W.; Bouckaert, L.; Dello Sterpaio, P.; Goswami, R.R.; Grinev, D.; Houston, A.; Hu, Y.; Liu, J.; et al. Observer-dependent variability of the thresholding step in the quantitative analysis of soil images and X-ray microtomography data. *Geoderma* **2010**, *157*, 51–63. [CrossRef]
24. Koyuncu, C.F.; Arslan, S.; Durmaz, I.; Cetin-Atalay, R.; Gunduz-Demir, C. Smart Markers for Watershed-Based Cell Segmentation. *PLoS ONE* **2012**, *7*, e48664. [CrossRef]
25. Brunke, O.; Brockdorf, K.; Drews, S.; Müller, B.; Donath, T.; Herzen, J.; Beckmann, F. Comparison between x-ray tube-based and synchrotron radiation-based uCT. In *Proceedings of SPIE 7078, Optical Engineering + Applications*; SPIE: Bellingham, WA, USA, 2008.
26. Kong, D.; Fonseca, J. Quantification of the morphology of shelly carbonate sands using 3D images. *Géotechnique* **2018**, *68*, 249–261. [CrossRef]
27. Chauhan, S.; Rühaak, W.; Anbergen, H.; Kabdenov, A.; Freise, M.; Wille, T.; Sass, I. Phase segmentation of X-ray computer tomography rock images using machine learning techniques: An accuracy and performance study. *Solid Earth* **2016**, *7*, 1125–1139. [CrossRef]
28. Chauhan, S.; Rühaak, W.; Khan, F.; Enzmann, F.; Mielke, P.; Kersten, M.; Sass, I. Processing of rock core microtomography images: Using seven different machine learning algorithms. *Comput. Geosci.* **2016**, *86*, 120–128. [CrossRef]
29. Krizhevsky, A.; Sutskever, I.; Hinton, G.E. ImageNet classification with deep convolutional neural networks. *Commun. ACM* **2017**, *60*, 84–90. [CrossRef]
30. Douarre, C.; Schielein, R.; Frindel, C.; Gerth, S.; Rousseau, D. Transfer Learning from Synthetic Data Applied to Soil–Root Segmentation in X-Ray Tomography Images. *J. Imaging* **2018**, *4*, 65. [CrossRef]
31. Karimpouli, S.; Tahmasebi, P. Segmentation of digital rock images using deep convolutional autoencoder networks. *Comput. Geosci.* **2019**, *126*, 142–150. [CrossRef]
32. Phan, J.; Ruspini, L.C.; Lindseth, F. Automatic segmentation tool for 3D digital rocks by deep learning. *Sci. Rep.* **2021**, *11*, 19123. [CrossRef]
33. Varfolomeev, I.; Yakimchuk, I.; Safonov, I. An Application of Deep Neural Networks for Segmentation of Microtomographic Images of Rock Samples. *Computers* **2019**, *8*, 72. [CrossRef]
34. Ronneberger, O.; Fischer, P.; Brox, T. U-Net: Convolutional Networks for Biomedical Image Segmentation. In *Medical Image Computing and Computer-Assisted Intervention—MICCAI 2015. Lecture Notes in Computer Science, Vol 9351*; Navab, N., Hornegger, J., Wells, W.M., Frangi, A.F., Eds.; Medical Image Computing and Computer-Assisted Intervention—MICCAI 2015; Springer International Publishing: Cham, Switzerland, 2015; pp. 234–241.
35. Brown, W.S. Physical Properties of Seawater. In *Springer Handbook of Ocean Engineering*; Dhanak, M.R., Xiros, N.I., Eds.; Springer: Cham, Switzerland, 2016.
36. Wadeson, N.; Basham, M.; Parsons, A.; Kazantsev, D.; Vo, N.T.; Schoonjans, T.; Pérez-Juárez, E.; Taylor, M.; Srikanth, N.; Nixon, D.; et al. *DiamondLightSource/Savu*, Version 2.4; 2019. Available online: https://doi.org/10.5281/zenodo.3541873 (accessed on 20 October 2020).
37. Atwood, R.C.; Bodey, A.J.; Price, S.W.T.; Basham, M.; Drakopoulos, M. A high-throughput system for high-quality tomographic reconstruction of large datasets at Diamond Light Source. *Philos. Trans. R. Soc. A* **2015**, *373*, 2369–2393. [CrossRef]
38. Wadeson, N.; Basham, M. Savu: A Python-based, MPI Framework for Simultaneous Processing of Multiple, N-dimensional, Large Tomography Datasets. *arXiv* **2016**, arXiv:1610.08015.

39. Paganin, D.; Mayo, S.C.; Gureyev, T.E.; Miller, P.R.; Wilkins, S.W. Simultaneous phase and amplitude extraction from a single defocused image of a homogeneous object. *J. Microsc.* **2002**, *206*, 33–40. [CrossRef]
40. Ramachandran, G.N.; Lakshminarayanan, A.V. Three-dimensional reconstruction from radiographs and electron micrographs: Application of convolutions instead of Fourier transforms. *Proc. Natl. Acad. Sci. USA* **1971**, *68*, 2236–2240. [CrossRef] [PubMed]
41. van Aarle, W.; Palenstijn, W.J.; Cant, J.; Janssens, E.; Bleichrodt, F.; Dabravolski, A.; De Beenhouwer, J.; Joost Batenburg, K.; Sijbers, J. Fast and flexible X-ray tomography using the ASTRA toolbox. *Opt. Express* **2016**, *24*, 25129–25147. [CrossRef] [PubMed]
42. Titarenko, V.; Bradley, R.; Martin, C.; Withers, P.; Titarenko, S. Regularization Methods for Inverse Problems in X-ray Tomography. In *Proceedings of the SPIE Optical Engineering + Applications*; SPIE: Bellingham, WA, USA, 2010.
43. Vo, N.T.; Drakopoulos, M.; Atwood, R.C.; Reinhard, C. Reliable method for calculating the center of rotation in parallel-beam tomography. *Opt. Express* **2014**, *22*, 19078–19086. [CrossRef] [PubMed]
44. Schindelin, J.; Arganda-Carreras, I.; Frise, E.; Kaynig, V.; Longair, M.; Pietzsch, T.; Preibisch, S.; Rueden, C.; Saalfeld, S.; Schmid, B.; et al. Fiji: An open-source platform for biological-image analysis. *Nat. Methods* **2012**, *9*, 676–682. [CrossRef] [PubMed]
45. Schneider, C.A.; Rasband, W.S.; Eliceiri, K.W. NIH Image to ImageJ: 25 years of image analysis. *Nat. Methods* **2012**, *9*, 671–675. [CrossRef]
46. Hsieh, J. *Computed Tomography: Principles, Design, Artifacts, and Recent Advances*, 3rd ed.; SPIE: Bellingham, WA, USA, 2015.
47. Kalender, W.A. *Computed Tomography. Fundamentals, System Technology, Image Quality, Applications*; Publicis Publishing: Erlangen, Germany, 2011.
48. Russakovsky, O.; Deng, J.; Su, H.; Krause, J.; Satheesh, S.; Ma, S.; Huang, Z.; Karpathy, A.; Khosla, A.; Bernstein, M.; et al. ImageNet Large Scale Visual Recognition Challenge. *Int. J. Comput. Vis.* **2015**, *115*, 211–252. [CrossRef]
49. Alvarez-Borges, F.J.; King, O.N.F.; Madhusudhan, B.N.; Ahmed, S.I. Tomography data of methane-bearing sand used to investigate U-Net segmentation methods [Dataset]. *Zenodo* **2021**. [CrossRef]
50. Paszke, A.; Gross, S.; Massa, F.; Lerer, A.; Bradbury, J.; Chanan, G.; Killeen, T.; Lin, Z.; Gimelshein, N.; Antiga, L.; et al. PyTorch: An Imperative Style, High-Performance Deep Learning Library. In *Advances in Neural Information Processing Systems 32*; Wallach, H., Larochelle, H., Beygelzimer, A., d'Alché-Buc, F., Fox, E., Garnett, R., Eds.; Curran Associates, Inc.: Red Hook, NY, USA, 2019; pp. 8024–8035.
51. Lee, K.; Zung, J.; Li, P.; Jain, V.; Seung, H.S. Superhuman Accuracy on the SNEMI3D Connectomics Challenge. *arXiv* **2017**, arXiv:1706.00120.
52. Wolny, A.; Cerrone, L.; Vijayan, A.; Tofanelli, R.; Barro, A.V.; Louveaux, M.; Wenzl, C.; Strauss, S.; Wilson-Sánchez, D.; Lymbouridou, R.; et al. Accurate and versatile 3D segmentation of plant tissues at cellular resolution. *eLife* **2020**, *9*, e57613. [CrossRef] [PubMed]
53. Pérez-García, F.; Sparks, R.; Ourselin, S. TorchIO: A Python library for efficient loading, preprocessing, augmentation and patch-based sampling of medical images in deep learning. *arXiv* **2020**, arXiv:2003.04696. [CrossRef]
54. Loshchilov, I.; Hutter, F. Decoupled Weight Decay Regularization. *arXiv* **2019**, arXiv:1711.05101.
55. Smith, L.N. Cyclical Learning Rates for Training Neural Networks. In Proceedings of the 2017 IEEE Winter Conference on Applications of Computer Vision, Santa Rosa, CA, USA, 24–31 March 2017; pp. 464–472.
56. King, O.N.F.; Alvarez-Borges, F.J. Gas Hydrate Segmentation Using U-Nets. Code Repository. 2021. Available online: https://github.com/DiamondLightSource/gas-hydrate-segmentation-unets (accessed on 14 April 2022).
57. He, K.; Zhang, X.; Ren, S.; Sun, J. Deep Residual Learning for Image Recognition. In Proceedings of the 2016 IEEE Conference on Computer Vision and Pattern Recognition, Las Vegas, NV, USA, 27–30 June 2016; pp. 770–778.
58. Howard, J.; Gugger, S. Fastai: A Layered API for Deep Learning. *Information* **2020**, *11*, 108. [CrossRef]
59. Tun, W.M.; Poologasundarampillai, G.; Bischof, H.; Nye, G.; King, O.N.F.; Basham, M.; Tokudome, Y.; Lewis, R.M.; Johnstone, E.D.; Brownbill, P.; et al. A massively multi-scale approach to characterising tissue architecture by synchrotron micro-CT applied to the human placenta. *bioRxiv* **2020**, *18*, 20210140. [CrossRef]
60. Smith, A.G.; Ørting, S. RootPainter 0.2.5. 2020. Available online: https://github.com/Abe404/root_painter (accessed on 15 March 2022).
61. Smith, A.G.; Han, E.; Petersen, J.; Olsen, N.A.F.; Giese, C.; Athmann, M.; Dresbøll, D.B.; Thorup-Kristensen, K. RootPainter: Deep Learning Segmentation of Biological Images with Corrective Annotation. *bioRxiv* **2020**. [CrossRef]
62. Smith, A.G.; Petersen, J.; Selvan, R.; Rasmussen, C.R. Segmentation of roots in soil with U-Net. *Plant Methods* **2020**, *16*, 13. [CrossRef]
63. Gonda, F.; Kaynig, V.; Jones, T.R.; Haehn, D.; Lichtman, J.W.; Parag, T.; Pfister, H. ICON: An interactive approach to train deep neural networks for segmentation of neuronal structures. In Proceedings of the IEEE 14th International Symposium on Biomedical Imaging, Melbourne, Australia, 18–21 April 2017.
64. Bradski, G. The OpenCV Library. 2000. Available online: https://www.drdobbs.com/open-source/the-opencv-library/184404319 (accessed on 15 April 2021).
65. Paris, S.; Kornprobst, P.; Tumblin, J.; Durand, F. Bilateral Filtering: Theory and Applications. *Found. Trends Comput. Graph. Vis.* **2009**, *4*, 1–73. [CrossRef]
66. Otsu, N. A Threshold Selection Method from Gray-Level Histograms. *IEEE Trans. Syst. Man Cybern.* **1979**, *9*, 62–66. [CrossRef]
67. van der Walt, S.; Schönberger, J.L.; Nunez-Iglesias, J.; Boulogne, F.; Warner, J.D.; Yager, N.; Gouillart, E.; Yu, T. Scikit-image: Image processing in Python. *PeerJ* **2014**, *2*, e453. [CrossRef]

68. Legland, D.; Arganda-Carreras, I.; Andrey, P. MorphoLibJ: Integrated library and plugins for mathematical morphology with ImageJ. *Bioinformatics* **2016**, *32*, 3532–3534. [CrossRef] [PubMed]
69. Karabağ, C.; Jones, M.L.; Peddie, C.J.; Weston, A.E.; Collinson, L.M.; Reyes-Aldasoro, C.C. Semantic segmentation of HeLa cells: An objective comparison between one traditional algorithm and four deep-learning architectures. *PLoS ONE* **2020**, *15*, e0230605. [CrossRef]
70. Kittler, J.; Illingworth, J. On threshold selection using clustering criteria. *IEEE Trans. Syst. Man Cybern.* **1985**, *5*, 652–655. [CrossRef]
71. Lee, S.U.; Yoon Chung, S.; Park, R.H. A comparative performance study of several global thresholding techniques for segmentation. *Comput. Vis. Graph. Image Process.* **1990**, *52*, 171–190. [CrossRef]
72. He, K.; Girshick, R.; Dollar, P. Rethinking ImageNet Pre-Training. In Proceedings of the IEEE/CVF International Conference on Computer Vision, Seoul, Republic of South Korea, 27 October–2 November 2019; pp. 4918–4927.
73. Matula, S.; Bát'ková, K.; Legese, W.L. Laboratory Performance of Five Selected Soil Moisture Sensors Applying Factory and Own Calibration Equations for Two Soil Media of Different Bulk Density and Salinity Levels. *Sensors* **2016**, *16*, 1912. [CrossRef] [PubMed]
74. Missimer, T.M.; Lopez, O.M. Laboratory Measurement of Total Porosity in Unconsolidated Quartz Sand by Two Integrated Methods. *J. Geol. Geophys.* **2018**, *7*, 1000448. [CrossRef]
75. Péron, H.; Hueckel, T.; Laloui, L. An Improved Volume Measurement for Determining Soil Water Retention Curves. *Geotech. Test. J.* **2007**, *30*, GTJ100167. [CrossRef]

methane

MDPI

Article

Enteric Methane Emission from Sheep Fed with Rhodes Grass Hay (*Chloris gayana*) Alone or Supplemented with Dried Distillers' Grains with Solubles

José Ignacio Gere [1,*], Mónica Feksa Frasson [2], Marisa Wawrzkiewicz [2], María Gabriela Fernández Pepi [2], María Laura Ramos [2], Ricardo Bualó [3], María Esperanza Cerón-Cucchi [3] and Gustavo Jaurena [2]

1 Engineering Research and Development Unit (UTNBA), National Technological University, Regional School of Buenos Aires, National Council for Scientific and Technical Research (CONICET), Ciudad Autónoma Buenos Aires C1179AAQ, Argentina

2 Animal Production Department, School of Agriculture, University of Buenos Aires (UBA), Ciudad Autónoma Buenos Aires C1417DSE, Argentina

3 Institute of Pathobiology, National Institute of Agricultural Technology (INTA), Hurlingham C1417AZE, Argentina

* Correspondence: jgere@frba.utn.edu.ar

Abstract: Livestock systems based on subtropical and tropical pastures are characterized by the low productivity of livestock due to the poor nutritional value of the forage (low nitrogen concentration and digestibility, and high fiber and lignin concentrations). These conditions lead to low productivity and, consequently, high absolute emissions of methane (CH_4) per unit of product. Dry distilled grains with solubles (DDGS) are the main by-product resulting from ethanol production, and they are characterized by their high-energy fibrous and protein content, thus becoming an option for the supplementation of low-quality forage. This research investigated the effects of dietary DDGS inclusion on dry matter digestibility (DMD) and enteric CH_4 emission. Eight adult sheep of 64 ± 8 kg live weight were used. The duration of the study was 54 days, divided into two periods (changeover design), which comprised a 17-day pre-experimental period and 10 days for experimental data collection. Animals were allocated to one of two treatments used: hay (H) as a control treatment, where animals were fed with Rhodes grass hay alone; and H + DDGS, where animals were fed with H supplemented with DDGS. CH_4 emissions were estimated using the sulfur hexafluoride (SF_6) tracer technique. Diets containing DDGS increased DMI by 22% ($p < 0.05$) and reduced daily CH_4 emissions by 24% (g/d), the CH_4 yield by 35% (g/kg DMI), and the average value of CH_4 energy per gross energy intake (Ym) by 44%, compared to the control treatment ($p < 0.05$). The experiment demonstrated that supplementation with DDGS in low-quality roughage reduced daily CH_4 emissions, yields, and Ym.

Keywords: sheep; agro-industry by-products; enteric methane emission; SF_6 tracer technique

1. Introduction

The agricultural sector faces the challenge of feeding a growing human population by 2050 [1], while meeting the social and environmental obligations of reducing greenhouse gas (GHG) emissions; the sector itself is responsible for producing 10 to 12% of the total global anthropogenic emissions [2]. However, the increased demand for protein-rich foods is leading to intensification of animal production and, hence, a likely increase in GHG emissions [2].

The production of enteric methane (CH_4) is a significant loss of the energy contained in feed [3], and although its persistence in the atmosphere is about 10 years, it has a warming effect about 28 times greater than that of CO_2 [4]. Of all animal production, enteric fermentation from ruminants is a major source of GHG emissions, accounting for 39% of all GHG emissions from the livestock sector [5] and between 11 and 13% of global CH_4 emissions [6]. Ruminant livestock represent one of the few sources that can be manipulated. This source is, furthermore, an attractive target for manipulation, since the reduction in CH_4 is usually associated with improved productivity [7–9].

Enteric CH_4 losses depend on several factors including feed intake, carbohydrate type, forage processing, lipid addition, and ruminal microbiota manipulation [3]. Feed intake is the most important predictor of CH_4 production [10,11]. Dry matter intake (DMI) is significantly and positively related to CH_4 production in adult sheep, with slopes ranging from 13.8 to 20.4 g CH_4 kg^{-1} DMI, and this relationship demonstrates that methanogenesis increases when more substrate is available for microbial fermentation in the rumen [11].

The quality of forage affects the activity of ruminal microbiota and CH_4 production in the rumen. Forage species, forage processing, the proportion of forage in the diet, and the source of the grain also influence CH_4 production in ruminants [8,12]. Total CH_4 production ($g \cdot d^{-1}$) tends to decrease as the protein content of feed increases, and it increases as the fiber content increases [8,13]

Ruminants produce proteins of high nutritional value, transforming fibrous forage resources that are not edible for humans. The symbiotic relationship with the rumen anaerobic microorganisms yields a high amount of hydrogen ions (H^+) that must be eliminated to keep the system functional. The main mechanism for the elimination of these ions is ruminal methanogenesis, which involves the reduction in carbon dioxide (CO_2) through the uptake of H^+ by Archaeobacteria [14]. Although there other metabolic pathways channel hydrogen, methanogenesis is the primary route of elimination.

The productivity of livestock in tropical and subtropical areas is usually low due to the low nutritional value of the available forages due to their highly lignified cell walls, low digestibility, and poor nitrogen content [15,16]. Under these conditions, supplementation with protein concentrates is an alternative having recognized productive benefits [17].

Dry distilled grains with solubles (DDGS) are a by-product from ethanol production, and due to their high energy and protein content, they can mostly replace grains [18] and, to a lesser extent, forages [19]. DDGS is a concentrate rich in crude protein (CP) (between 27% and 30%) and lipids (between 8% and 11%), and it also contributes with fiber, phosphorus, and lower concentrations of starch compared to the grain from which it is derived [20–22].

The use of industrial by-products of animal production systems leads to a reduction in the environmental impact and may also cause a reduction in the cost of waste treatment. Therefore, these reductions generate an economic benefit in terms of the added value given to by-products and waste [23]. In agricultural systems, it is necessary to modify the resource inputs and flows by increasing on-farm and farm-to-farm recycling, by redirecting current outputs into inputs for other production systems, and by reducing input costs and recovering income from resources that would otherwise be wasted and could harm the environment [24].

The aim of this study was to assess the effect of adding DDGS to Rhodes grass hay on dry matter digestibility and enteric CH_4 emissions from sheep.

2. Results

Table 1 shows the chemical composition of diet for both treatments. The inclusion of DDGS improved the quality of feed offered by increasing the CP (from 74 to 149 $g \cdot kg^{-1}$ DM), EE (from 15 to 54 $g \cdot kg^{-1}$ DM), WSC (from 40 to 49 $g \cdot kg^{-1}$ DM), starch (from 78 to 83), and DMD (from 310 to 450 $g \cdot kg^{-1}$ DM), and reduced the NDF (from 737 to 616 $g \cdot kg^{-1}$ DM), the ADF (from 401 to 293 $g \cdot kg^{-1}$ DM), and the lignin content (72 to 51 $g \cdot kg^{-1}$ DM) for H and H + DDGS, respectively.

Table 1. Feedstuffs chemical composition (g/kg dry matter, except stated otherwise).

Chemical Fraction	Feed		
	Hay	DDGS	Hay + DDGS
Dry matter (g·kg^{-1} as fed)	806	796	806
Ash	136	49	118
Crude Protein	74	285	149
Neutral detergent fibre	737	440	616
Acid detergent fibre	401	120	293
Lignin	72	22	51
Ether extract	15	120	54
Water soluble carbohydrates	40	71	49
Starch	78	96	83
Dry matter digestibility	310	-	450
Gross energy (MJ·kg^{-1})	17	21	19

The results obtained when evaluating the daily CH_4 emissions, the DMI, and the emissions related to DMI are shown in Table 2. Animals on the H + DDGS treatment presented a significantly higher DMI (827 vs. 679 g·d^{-1}) and lower CH_4 emissions (16 vs. 21 g·d^{-1}) than those on the H treatment and, consequently, they emitted less CH_4 when evaluating the CH_4 yield (20 vs. 31 g·kg^{-1} DMI). The H + DDGS-fed animals presented a CH_4 energy loss through eructation (Ym) of 5.7%, and the H-fed animals showed 10.1%.

Table 2. Dry matter intake and enteric CH_4 production of sheep fed hay alone or supplemented with DDGS.

	Treatments		SEM [1]	*p* Value
	H	H + DDGS		
Dry matter intake (g·d^{-1})				
Hay	679	535	65	0.054
DDGS	0	292	-	-
Total	679	827	69	0.035
Total (% liveweight)	1.2	1.5	0.14	0.049
CH_4 emission				
CH_4 (g·d^{-1})	21	16	1.1	0.014
CH_4 (g·kg^{-1} dry matter intake)	31	20	1.9	0.005
Ym (%) [2]	10.1	5.7	0.6	0.002

[1] Standard error of the mean. [2] Energy loss through eructation.

3. Materials and Methods

3.1. Experimental Treatments, Study Location and Animal Procedures

The experiment was carried out in the Animal Production Department of the School of Agriculture (University of Buenos Aires; Buenos Aires, Argentina), in conjunction with the Rumen Microbiology Laboratory (National Institute of Agricultural Technology; Hurlingham, Argentina) and the National Technological University (Regional School of Buenos Aires; Buenos Aires, Argentina). The experimental protocols, procedures, and the care of the animals were approved by the Ethics and Animal Welfare Committee (N° 5229/2017) of the University of Buenos Aires, Argentina.

Eight adult Friesian sheep (*Ovis aries*) having a 64 ± 8 kg live weight were used, of which four had permanent ruminal cannulas. The duration of the study was 54 days, divided into two periods (changeover design), which comprised a 17-day pre-experimental period and 10 days for experimental data collection. The pre-experimental phase entailed the adaptation of the animals to the canisters, the placement, and monitoring of the permeation tubes, and adaptation to the experimental diet. The experimental stage involved daily collection of feces, urine, and feed intake. The measurement of enteric CH_4 emissions was carried out in the last 5 days of this period.

Two treatments were used: (1) hay (H), where animals were fed with Rhodes grass hay alone; and (2) Hay + DDGS (H + DDGS), where animals were fed with Rhodes grass hay with added DDGS (ratio of 64:36 on a dry matter basis; Table 1). Cannulated animals were housed in individual pens, and the remainder of the animals were housed in metabolic cages. Both groups were fed ad libitum once a day (8 a.m.) with free access to water.

The voluntary DMI of all animals was calculated as the difference between the offered and rejected ingredients for each period (pool samples of daily collected aliquots). At the end of each period, the pool sample was frozen until subsequent drying and preparation for chemical analysis. Total collections of feces and urine from animals in metabolic cages were conducted to compute the energy and nitrogen balances, which will be reported in a subsequent article.

3.2. Measurement of Enteric Methane Emissions

For the quantification of enteric CH_4, the sulfur hexafluoride (SF_6) tracer gas technique proposed by Johnson et al. was used [3]. At the beginning of the acclimatization period, the sheep were orally dosed with brass permeation tubes containing SF_6 (ca. 1 g of SF_6 per tube; mean permeation rate 2.26 ± 0.56 mg·d^{-1}), which were prepared at the Institute of Pathobiology (INTA) 2 months before the experiment and calibrated for 4 weeks. The sampling period of collected exhaled air was 5 consecutive days. The sample system consisted of a polyvinyl chloride (PVC) yoke-shape collection device (1.25 L volume) with a sample flow regulated by a capillary system. The concentrations of CH_4 and SF_6 were analyzed using a gas chromatograph (Perkin Elmer 600, Kansas City, USA), as described by Gere et al. [25].

3.3. Chemical Analysis

All procedures were adjusted according to the standardized protocols of the Program for the Improvement of the Evaluation of Forages and Feeds [26]. The feed, refusals, and feces samples were dried (65 °C, 48 h) and ground (1 mm; Willey-type mill) before characterization. All results are reported on a dry matter (DM) basis (105 °C for 4 h, AOAC, 1991; No. 976.63). The ash content was determined after complete ignition at 500 °C for 4 h (AOAC, 1995; No. 942.05). The content of Pro-Nitro® protein (Selecta J.P., Barcelona, Spain) and the ether extract was determined in Soxhlet with petroleum ether [27]. The neutral detergent fiber (NDF) was reported ash-free and determined according to the Van Soest et al. methodology without sodium sulfite and using thermostable amylase [28]. Subsequently, the NDF was corrected for ashes (aNDFmo). The acid detergent fiber (ADF_{MO}) and the lignin contents (LDA_{MO}) were reported ash-free according to Goering and Van Soest [29] and Van Soest et al. [28] and determined using ANKOM® equipment (Model 220). The total starch content was measured using the AA/AMG Megazyme enzyme kit (Megazyme Ltd., Neogen, Ireland). The content of water-soluble carbohydrates (WSC) was also determined by colorimetry using the Antrona method [30]. The gross energy (GE) was determined using a bomb calorimeter (PARR 1261, Parr Instrument Company, Moline, IL, USA).

3.4. Statistical Analysis

Results were analyzed according to a double Latin square experimental design (one Latin square with cannulated animals, and the other with non-cannulated animals; feed, period), using proc Mixed (SAS Version 8.0, SAS Institute Inc. Cary, NC, USA). Differences were declared significant when $p < 0.05$.

4. Discussion

Supplementing diets with DGGS as a source of protein increased digestibility and total DMI, as previously found by McCollum et al. [31], Beaty et al. [32], and Mathis et al. [33]. Hence, the ration with DDGS increased the total DMI by 22% compared to hay alone (Table 2; $p = 0.035$), increasing the DMI from 1.2% to 1.5% of body weight ($p = 0.049$). This result agrees with the results previously reported by Winterholler et al. [34], who used low quality hay (PC = 5.9%) and a range of DDGS supplementation levels (0.3, 0.75, 1.20, and 1.65% PV) in a beef steer diet and observed a linear increase in total DMI; and Morris et al. [35], who found in low-quality forage a substitution rate of 0.32 kg for every additional kilogram of DDGS fed. Schauer et al. [36] and Felix et al. [37] reported that lambs could be fed with 60% DDGS (DM basis), without affecting DMI and animal performance, and the optimum dietary inclusion of DDGS for lambs occurred at 20% of the DM [37].

The average CH_4 emissions (21 and 16 $g \cdot d^{-1}$ for H and H + DDGS, respectively) were in the range of reported values of a database of individual sheep records from CH_4 emission studies conducted in the Latin America and Caribbean (LAC) region (219 individual sheep records from 11 studies) [11]. The summary report for mature animals (n = 79, mean BW of 52.4 kg, DMI 1.35 $kg \cdot d^{-1}$) showed values between 14 and 57 $g \cdot d^{-1}$, with a mean value of 30 $g \cdot d^{-1}$ [11].

Diets containing DDGS reduced the CH_4 yield by 35% ($g \cdot kg^{-1}$ DMI, $p = 0.005$; Table 2) compared to the control treatment (hay alone). The decrease in CH_4 emissions agreed with the increase in DMI and DMD. Similarly, McGinn et al. observed a reduction in CH_4 yield when Hereford steers were supplemented with DDGS receiving a base diet of barley silage [38].

In addition, it should be noted that the H + DDGS ration was 54 g $EE \cdot kg^{-1}$ DM (Table 1), which could have contributed to the reduction in CH_4 emissions. It has been noted that EE can negatively affect the emissions of CH_4. The lipid content of DDGS (120 g $EE \cdot kg^{-1}$ DM; Table 1) increased the crude fat content from 15 to 54 g $EE \cdot kg^{-1}$ DM for H and H + DDGS, respectively. As result of a meta-analysis, Beauchemin et al. [39] concluded that, for each 1% of lipid added in the diet, there was a 5.6% reduction in the production of enteric CH_4 ($g \cdot kg^{-1}$ DMI). Benchaar et al. [40] worked with dairy cattle fed increasing levels of DDGS in the diet (10, 20, and 30% of the DM replacing flaked corn and soybean meal) and found that enteric CH_4 yield decreased by 0.5 $g \cdot kg^{-1}$ DMI (0.5, 0.4, and 0.7 for 10%, 20%, and 30% of the DM, respectively). The reduction observed in our experiment was higher, and closer to the prediction reported by Beauchemin et al. [39], who signaled considerable variation in the CH_4 reductions observed among fat sources.

Several studies have reported decreases in enteric CH_4 emissions when cattle diets were supplemented with unprotected fat [41–43]. It has been argued that a decrease in CH_4 emissions is due to the reduction in organic matter fermented in the rumen and by the toxic effects on cellulolytic bacteria, methanogen activity, and number of protozoa [39,44].

The production of alcohol and carbon dioxide from maize grain requires starch removal from the grain; hence, the remaining nutrient concentration in the DDGS increases approximately three-fold [20]. Several factors such as the original grain quality and industrial process, among others, can influence the nutritional and physical properties of DDGS, which is usually considered a highly variable by-product (e.g., residual starch, WSC, lipids). Aside from the lipid content, starch and WSC can also contribute to the reduction in CH_4 production [5]. The literature reports average values for Ym of 5.4% of gross energy intake (GEI) for grazing sheep [45]. For growing lambs, Savian et al. [46] reported an average value of 7.3% (grazing ryegrass), and Amaral et al. [47] reported an average value of 5% (grazing pearl millet). The summary report of Congio et al. [11] showed values from 4.4% to 11.6%, with a mean value of 7%. The average values of Ym in this study were 10.1% and 5.7% for the H and H + DDGS treatments, respectively, agreeing with the results found in the literature (Table 2).

5. Conclusions

These results showed that supplementation with DDGS on Rhodes grass hay reduced CH_4 emissions from sheep. This effect was associated with a greater DMI and higher DMD and EE concentration in the diet. These results suggest industrial by-products as supplements for low-quality diets may be a promising CH_4 emission mitigation strategy.

Author Contributions: Conceptualization, J.I.G., M.E.C.-C. and G.J.; methodology, J.I.G., M.F.F., M.W., M.G.F.P., R.B., M.E.C.-C. and G.J.; software, M.F.F. and J.I.G.; validation, J.I.G., M.E.C.-C. and G.J.; formal analysis, J.I.G., M.F.F., M.E.C.-C., M.L.R., and G.J.; investigation, J.I.G., M.F.F., M.E.C.-C. and G.J.; resources, J.I.G., M.F.F., M.W., M.G.F.P., R.B., M.E.C.-C. and G.J.; data curation, J.I.G., M.F.F., M.W., M.G.F.P., R.B., M.E.C.-C. and G.J.; writing—original draft preparation, J.I.G. and M.F.F.; writing—review and editing, J.I.G., M.F.F., M.E.C.-C. and G.J.; visualization, J.I.G., M.E.C.-C. and G.J.; supervision, J.I.G., M.E.C.-C. and G.J.; project administration, J.I.G., M.E.C.-C. and G.J.; funding acquisition, J.I.G., M.E.C.-C. and G.J. All authors have read and agreed to the published version of the manuscript.

Funding: The Project was funded by UBACyT 2018 GC 651BA; PID UTN MSIFNBA0005518, PICT-2019-1647 and INTA PNNAT 1128023.

Institutional Review Board Statement: The experimental protocols, procedures, and the care of the animals were approved by the Ethics and Animal Welfare Committee (N° 5229/2017) of the University of Buenos Aires, Argentina.

Informed Consent Statement: Not applicable.

Data Availability Statement: Not applicable.

Acknowledgments: The authors would like to thank the technicians supports of Animal Production Department, School of Agriculture, University of Buenos Aires and the English Language editing and review services supplied by the Academic Translation Centre of the UTN Buenos Aires. We also thank Dra. Gabriela Posse Beaulieu for her valuable contribution in funding acquisition for enteric CH_4 emission measurements.

Conflicts of Interest: The authors declare no conflict of interest. The founding sponsors had no role in the design of the study; in the collection, analyses, or interpretation of data; in the writing of the manuscript; and in the decision to publish the results.

References

1. FAO/WHO Expert Committee on Food Additives. Meeting. Joint FAO/WHO Expert Committee on Food Additives, & Meeting Staff. Compendium of Food Additive Specifications: Joint FAO/WHO Expert Committee on Food Additives: 67th Meeting. Food and Agriculture Organization. 2006. Available online: http://www.fao.org (accessed on 24 August 2022).
2. Solomon, S.; Qin, D.; Manning, M.; Chen, Z.; Marquis, M.; Averyt, K.; Tignor, M.; Miller, H. IPCC Fourth Assessment Report (AR4). In *Climate Change*; IPCC: Geneva, Switzerland, 2007; pp. 133–171.
3. Johnson, K.; Huyler, M.; Westberg, H.; Lamb, B.; Zlmmerman, P. Measurement of Methane Emissions from Ruminant Livestock Using a SF6 Tracer Technique. *Environ. Sci. Technol.* **1994**, *28*, 359–362. [CrossRef] [PubMed]
4. Pachauri, R.K.; Allen, M.R.; Barros, V.R.; Broome, J.; Cramer, W.; Christ, R.; Church, J.A.; Clarke, L.; Dahe, Q.; Dasgupta, P. Climate Change 2014: Synthesis Report. In *Contribution of Working Groups I, II and III to the Fifth Assessment Report of the Intergovernmental Panel on Climate Change*; IPCC: Geneva, Switzerland, 2014.
5. Gerber, P.; Steinfeld, H.; Henderson, B.; Mottet, A.; Opio, C.; Dijkman, J.; Falcucci, A.; Tempio, G. *Tackling Climate Change through Livestock—A Global Assessment of Emissions and Mitigation Opportunities*; Food and Agriculture Organization: Rome, Italy, 2013.
6. Beauchemin, K.A.; McAllister, T.A.; McGinn, S.M. Dietary mitigation of enteric methane from cattle. *CAB Rev. Perspect. Agric. Vet. Sci. Nutr. Nat. Resour.* **2009**, *4*, 1–18.
7. Leng, R.A. Quantitative ruminant nutrition—A green science. *Aust. J. Agric. Res.* **1993**, *44*, 363–380.
8. Shibata, M.; Terada, F. Factors affecting methane production and mitigation in ruminants. *Anim. Sci. J.* **2010**, *81*, 2–10. [PubMed]
9. Min, B.-R.; Lee, S.; Jung, H.; Miller, D.N.; Chen, R. Enteric Methane Emissions and Animal Performance in Dairy and Beef Cattle Production: Strategies, Opportunities, and Impact of Reducing Emissions. *Animals* **2022**, *12*, 948.
10. Patra, A.K.; Lalhriatpuii, M.; Debnath, B.C. Predicting enteric methane emission in sheep using linear and non-linear statistical models from dietary variables. *Anim. Prod. Sci.* **2016**, *56*, 574–584.
11. Congio, G.F.; Bannink, A.; Mayorga, O.L.; Rodrigues, J.P.; Bougouin, A.; Kebreab, E.; Carvalho, C.F.; Abdalla, L.; Monteiro, L.G.; Hristov, A.N.; et al. Prediction of enteric methane production and yield in sheep using a Latin America and Caribbean database. *Livest. Sci.* **2022**. [CrossRef]
12. Savin, K.W.; Moate, P.J.; Williams, S.R.O.; Bath, C.; Hemsworth, J.; Wang, J.; Ram, D.; Zawadzki, J.; Rochfort, S.; Cocks, B.G. Dietary wheat and reduced methane yield are linked to rumen microbiome changes in dairy cows. *PLoS ONE* **2022**, *17*, e0268157.
13. Tyagi, N.; Mishra, D.B.; Vinay, V.V.; Kumar, S. Feasible Strategies for Enteric Methane Mitigation from Dairy Animals. In *Animal Manure. Soil Biology*; Mahajan, S., Varma, A., Eds.; Springer: Cham, Germany, 2022; Volume 64, pp. 335–354.
14. Wolin, M.; Miller, T.; Stewart, C. Microbe-microbe interactions. In *The Rumen Microbial Ecosystem*, 1st ed.; Hobson, P.N., Stewart, C.S., Eds.; Springer: Dordrecht, The Netherlands, 1997; pp. 467–491.

15. Goel, G.; Makkar, H.P. Methane mitigation from ruminants using tannins and saponins. *Trop. Anim. Health Prod.* **2012**, *44*, 729–739.
16. Lee, M.A.; Davis, A.P.; Chagunda, M.G.G.; Manning, P. Forage quality declines with rising temperatures, with implications for livestock production and methane emissions. *Biogeosciences* **2017**, *14*, 1403–1417.
17. Cooke, R.F.; Caigle, C.L.; Moriel, P.; Smith, S.B.; Tedeschi, L.O.; Vendramini, J.M.B. Cattle adapted to tropical and subtropical environments: Social, nutritional, and carcass quality considerations. *J. Anim. Sci.* **2020**, *98*, 1–20.
18. Klopfenstein, T.J.; Erickson, G.E.; Bremer, V.R. Board-invited review: Use of distillers by-products in the beef cattle feeding industry. *J. Anim. Sci.* **2008**, *86*, 1223–1231. [PubMed]
19. Li, Y.L.; McAllister, T.A.; Beauchemin, K.A.; He, M.L.; McKinnon, J.J.; Yang, W.Z. Substitution of wheat dried distillers grains with solubles for barley grain or barley silage in feedlot cattle diets: Intake, digestibility, and ruminal fermentation. *J. Anim. Sci.* **2011**, *89*, 2491–2501. [PubMed]
20. Spiehs, M.J.; Whitney, M.H.; Shurson, G.C. Nutrient database for distiller's dried grains with solubles produced from new ethanol plants in Minnesota and South Dakota. *J. Anim. Sci.* **2002**, *80*, 2639–2645. [PubMed]
21. National Academies of Sciences, Engineering, and Medicine. *Nutrient Requirements of Beef Cattle*, 8th ed.; National Academies Press: Washington, DC, USA, 2016; ISBN 978-0-309-31702-3.
22. Paulus Compart, D.M.; Carlson, A.M.; Crawford, G.I.; Fink, R.C.; Diez-Gonzalez, F.; DiCostanzo, A.; Shurson, G.C. Presence and biological activity of antibiotics used in fuel ethanol and corn co-product production. *J. Anim. Sci.* **2013**, *91*, 2395–2404.
23. Taheripour, F.; Hertel, T.W.; Tyner, W.E.; Beckman, J.F.; Birur, D.K. Biofuels and their by-products: Global economic and environmental implications. *Biomass Bioenergy* **2010**, *34*, 278–289. [CrossRef]
24. Morton, L.W.; Shea, E. Frontier: Beyond Productivity—Recreating the Circles of Life to Deliver Multiple Benefits with Circular Systems. *J. Agric. Saf. Health* **2022**, *65*, 411–418. [CrossRef]
25. Gere, J.I.; Bualó, R.A.; Perini, A.L.; Arias, R.D.; Ortega, F.M.; Wulff, A.E.; Berra, G. Methane emission factors for beef cows in Argentina: Effect of diet quality. *N. Z. J. Agri. Res.* **2019**, *64*, 260–268. [CrossRef]
26. Jaurena, G.; Wawrzkiewicz, M. Programa para el Mejoramiento de la Evaluación de Forrajes y Alimentos (PROMEFA). In *Guía de procedimientos, Centro de Investigación y Servicios en Nutrición Animal*; Facultad de Agronomía-Universidad de Buenos Aires: Buenos Aires, Argentina, 2021.
27. Helrich, K. Association of Official Analytical Chemists. In *Official Methods of Analysis, Association of the Official Analytical Chemists*, 15th ed.; Association of Official Analytical Chemists: Arlington, VA, USA, 1990.
28. Van Soest, P.J.; Robertson, J.B.; Lewis, B.A. Methods for dietary fiber, neutral detergent fiber, and nonstarch polysaccharides in relation to animal nutrition. *J. Dairy Sci.* **1991**, *74*, 3583–3597.
29. Goering, H.R.; van Soest, P.J. *Forage Fiber Analyses; Agricultural Handbook No. 379*; United States Department of Agricultum: Washington, DC, USA, 1970.
30. Yemm, E.W.; Willis, A.J. The estimation of carbohydrates in plant extracts by anthrone. *Biochem. J.* **1954**, *57*, 508–514. [CrossRef]
31. Mccollum, F.T.; Galyean, M.L.; Krysl, L.J.; Wallace, J.D. Cattle Grazing Blue Grama Rangeland I. Seasonal Diets and Rumen Fermentation. *J. Range Manag.* **1985**, *38*, 539–543. [CrossRef]
32. Beaty, J.L.; Cochran, R.C.; Lintzenich, B.A.; Vanzant, E.S.; Morrill, J.L.; Brandt, R.T., Jr.; Johnson, D.E. Effect of frequency of supplementation and protein concentration in supplements on performance and digestion characteristics of beef cattle consuming low-quality forages. *J. Anim. Sci.* **1994**, *72*, 2475–2486. [CrossRef] [PubMed]
33. Mathis, C.P.; Cochran, R.C.; Stokka, G.L.; Heldt, J.S.; Woods, B.C.; Olson, K.C. Impacts of increasing amounts of supplemental soybean meal on intake and digestion by beef steers and performance by beef cows consuming low-quality tallgrass prairie forage. *J. Anim. Sci.* **1999**, *77*, 3156–3162. [CrossRef]
34. Winterholler, S.J.; Holland, B.P.; McMurphy, C.P.; Krehbiel, C.R.; Horn, G.W.; y Lalman, D.L. Use of dried distillers grains in preconditioning programs for weaned beef calves and subsequent impact on wheat pasture, feedlot, and carcass performance. *Prof. Anim. Sci.* **2009**, *25*, 722–730. [CrossRef]
35. Morris, S.E.; Klopfenstein, T.J.; Adams, D.C.; Erickson, G.E.; VanderPol, K.J. The effects of dried distillers grains on heifers consuming low or high quality forage. *Nebraska Beef Rep.* **2005**, 18–20.
36. Schauer, C.S.; Stamm, M.M.; Maddock, T.D.; Berg, P.B. Feeding of DDGS in lamb rations. *Sheep Goat Res. J.* **2008**, *23*, 15–19.
37. Felix, T.L.; Zerby, H.N.; Moeller, S.J.; Loerch, S.C. Effects of increasing dried distillers grains with soluble on performance, carcass characteristics, and digestibility of feedlot lambs. *J. Anim. Sci.* **2012**, *90*, 1356–1363. [CrossRef]
38. McGinn, S.M.; Beauchemin, K.A.; Flesch, T.K.; Coates, T. Performance of a dispersion model to estimate methane loss from cattle in pens. *J. Environ. Qual.* **2009**, *38*, 1796–1802. [CrossRef]
39. Beauchemin, K.A.; Kreuzer, M.; O'Mara, F.; McAllister, T.A. Nutritional management for enteric methane abatement: A review. *Aust. J. Exp. Agric.* **2008**, *48*, 21–27. [CrossRef]
40. Benchaar, C.; Hassanat, F.; Gervais, R.; Chouinard, P.Y.; Julien, C.; Petit, H.V. Effects of increasing amounts of corn-dried distillers' grains with solubles in dairy cow diets on methane production, ruminal fermentation, digestion, N balance, and milk production. *J. Dairy Sci.* **2013**, *96*, 2413–2427. [CrossRef]
41. Beauchemin, K.A.; McGinn, S.M.; Benchaar, C.; Holtshausen, L. Crushed sunflower, flax, or canola seeds in lactating dairy cow diets: Effects on methane production, rumen fermentation, and milk production. *J. Dairy Sci.* **2009**, *92*, 2118–2127. [CrossRef] [PubMed]

42. Grainger, C.; Williams, R.; Clarke, T.; Wright, A.-D.; Eckard, R. Supplementation with whole cottonseed causes long-term reduction of methane emissions from lactating dairy cows offered a forage and cereal grain diet. *J. Dairy Sci.* **2010**, *93*, 2612–2619. [CrossRef] [PubMed]
43. Chung, Y.H.; He, M.L.; McGinn, S.M.; McAllister, T.A.; Beauchemin, K.A. Linseed suppresses enteric methane emissions from cattle fed barley silage, but not from those fed grass hay. *Anim. Feed Sci. Technol.* **2011**, *166–167*, 321–329. [CrossRef]
44. Johnson, D.E.; Johnson, K.A. Methane emissions from cattle. *J. Anim. Sci.* **1995**, *73*, 2483–2492. [CrossRef]
45. Lassey, K.R. Livestock methane emission: From the individual grazing animal through national inventories to the global methane cycle. *Agric. For. Meteorol.* **2007**, *142*, 120–132. [CrossRef]
46. Amaral, G.A.; David, D.B.; Gere, J.I.; Savian, J.V.; Kohmann, M.M.; Nadin, L.B.; Sánchez, F.; Bayer, C.; Carvalho, P.C. Methane emissions from sheep grazing pearl millet (Pennisetum americanum (L.) Leek) swards fertilized with increasing nitrogen levels. *Small Rumin. Res.* **2016**, *141*, 118–123. [CrossRef]
47. Savian, J.V.; Neto, A.B.; de David, D.B.; Bremm, C.; Schons, R.M.T.; Genro, T.C.M.; do Amaral, G.A.; Gere, J.; McManus, C.M.; Bayer, C.; et al. Grazing intensity and stocking methods on animal production and methane emission by grazing sheep. Implications for integrated crop-livestock system. *Agric. Ecosyst. Environ.* **2014**, *190*, 112–119. [CrossRef]

methane

MDPI

Review

Opportunities and Hurdles to the Adoption and Enhanced Efficacy of Feed Additives towards Pronounced Mitigation of Enteric Methane Emissions from Ruminant Livestock

Emilio M. Ungerfeld

Centro Regional de Investigación Carillanca, Instituto de Investigaciones Agropecuarias INIA, Camino Cajón a Vilcún km 10, Vilcún 4880000, Chile; emilio.ungerfeld@inia.cl; Tel.: +56-452297296

Abstract: This paper analyzes the mitigation of enteric methane (CH_4) emissions from ruminants with the use of feed additives inhibiting rumen methanogenesis to limit the global temperature increase to 1.5 °C. A mathematical simulation conducted herein predicted that pronounced inhibition of rumen methanogenesis with pure chemicals or bromoform-containing algae with an efficacy higher than that obtained in most studies can be important to limiting global temperature increase by 2050 to 1.5 °C but will likely need to be accompanied by improved production efficiency and other mitigation measures. Currently, the most important limitations to the adoption of antimethanogenic feed additives are increased feeding cost without a consistent return in production efficiency and achieving sustained delivery of inhibitors to grazing animals, especially in extensive systems. Economic incentives could be applied in some countries to favor adoption of inhibitors. Changes in rumen microbial and whole animal metabolism caused by inhibiting methanogenesis could potentially be used to make the methanogenesis inhibition intervention cost-effective, although research in this direction is unlikely to yield results in the short term. Future research directions to maximize the adoption and efficacy of inhibitors of methanogenesis are examined.

Keywords: enteric methane; ruminants; mitigation; rumen; methanogenesis inhibition; feed additives; adoption; cost effectiveness

1. Enteric Methane Emissions and Climate Change

There is consensus that, in comparison to 2 °C or even higher levels of global temperature increase, limiting global temperature increase to 1.5 °C will diminish the frequency and severity of extreme climate events in the next decades [1]. Methane (CH_4) atmospheric concentration has doubled since industrial times and is currently second to carbon dioxide (CO_2) in causing global warming [2]. In addition to reaching net zero emissions of CO_2, achieving a strong, rapid, and sustained decrease in CH_4 emissions is key to rapidly limiting global warming [3]. This is largely due to CH_4's relatively high global warming potential (28 times greater than CO_2 in a 100-year period) and relatively short life (9.25 ± 0.6 years) and perturbation time (12.4 ± 1.4 years) in the atmosphere [4]. Other benefits of decreasing CH_4 concentration in the atmosphere include preventing premature death due to ground-level ozone pollution and increasing crop yields [2].

As part of the overall mitigation in the emissions of greenhouse gases (GHG) to limit global warming to 1.5 °C in this century, it is estimated that global anthropogenic CH_4 emissions must be reduced by 40 to 45% by 2030 from 2015 levels [2]. On the other hand, past and recent trends indicate continuous growth in the emissions of CH_4, with a recent acceleration and projected increases in the atmospheric concentration of CH_4 under the current scenario [2,4–6]. Agriculture is a major source of short-term global warming through its emissions of CH_4 [4]. Enteric CH_4 emitted by domestic ruminants is the main source of agricultural CH_4 and accounts for about 30% of total CH_4 emissions from human activities [2,7]. Emissions of CH_4 by livestock increased by 51.4% between 1961 and 2018 [7].

The necessary decrease in enteric CH_4 emissions between 2020 and 2030 across various socioeconomic scenarios and climate models, compatible with a maximal 1.5 °C increase in global temperature, was estimated to be 20% on average [2]. Decreasing enteric CH_4 emissions is, therefore, important as part of the effort to decrease the anthropogenic emissions of GHG. The objectives of this paper are to critically examine through a mathematical simulation the possibilities of decreasing enteric CH_4 emissions through sustainable intensification of ruminant agriculture and the use of feed additive inhibitors of methanogenesis as the most potent strategy for enteric CH_4 mitigation and to analyze the opportunities and barriers to widespread adoption of inhibitors of methanogenesis for pronounced mitigation of enteric CH_4 emissions.

2. Intensification, Productivity, and Enteric Methane Emissions

Intensifying ruminant production increases the feed intake and productivity of the individual animal. Feed intake is the main driver of CH_4 production [8]. Increased feed intake resulting from improved feed availability and quality thus results in greater daily CH_4 emissions per animal. On the other hand, as animal productivity increases, a lesser proportion of dry matter intake (DMI) and of CH_4 emitted by an animal is associated with the animal's maintenance requirements, which has been called the "dilution of maintenance" effect. The result is a decrease in CH_4 emitted per unit of milk [9] or meat [10] produced or CH_4 intensity. There are also other animal management and feeding practices that also improve animal productivity and decrease CH_4 intensity, such as reducing herd size to increase individual productivity, reducing mortality and morbidity, decreasing age at slaughter, and improving fertility [11].

Improvements in production efficiency between the 2000–2004 and 2014–2018 quinquennials led to declines in CH_4 intensity of meat and milk from dairy cattle, buffalo, sheep, and goat protein in most regions in the world, although this was more variable for beef. Despite the decreases in CH_4 intensity, total CH_4 emitted globally by ruminants increased in the same period of time [7]. Due to the forecasted increase in production of animal products, Chang et al. [7] projected a global increase in total emissions of livestock CH_4 (including pigs and poultry) of between 51 and 54% by 2050 relative to 2012 assuming constant CH_4 intensities. With decreasing CH_4 intensities due to improved production efficiency following past trends, total global emissions of CH_4 from livestock were estimated to increase less, by 15 to 21%, between 2012 and 2050 [7]. A similar analysis for wool production in Western Australia also revealed a relationship between increased animal productivity, mostly attributed to improvements in reproductive performance, and decreased CH_4 intensity, along with increased total emissions of CH_4 [12]. Therefore, whilst production intensification and resulting improvements in animal productivity and feed efficiency can ameliorate livestock CH_4 emissions relative to a scenario with constant CH_4 intensity, total CH_4 emissions from livestock will likely continue rising, as a result of the increases in animal production that are, in turn, driven by the increases in human population and per capita consumption of animal products, especially in developing countries [13,14].

It has been estimated that agricultural emissions of CH_4 must diminish between 24 and 47% by 2050 relative to a 2010 baseline in order to contain the global temperature increase to 1.5 °C [15]. Given that the main source of agricultural CH_4 is livestock [16], it is reasonable to assume that enteric CH_4 will also need to be decreased by similar percentages between 2010 and 2050. In the same period, the consumption of bovine and ovine meat and dairy products is expected to expand by 58, 78, and 58%, respectively [13]. It follows that, in order to decrease enteric CH_4 emissions by 24% by 2050 relative to 2010 levels, global CH_4 intensity of beef, lamb, and milk production would have to decrease by 52, 57, and 52%, respectively, in relation to its 2010 levels. Likewise, decreasing enteric CH_4 emissions by 47% between 2010 and 2050 would require decreasing global CH_4 emissions intensity of beef, lamb, and milk production by 66, 70, and 66%, respectively (calculations not shown).

The same as with CH_4, intensifying animal production and improving animal productivity also decreases the emissions intensity of carbon dioxide equivalents (CO_2e; the sum of the main three GHG CO_2, CH_4, and nitrous oxide (N_2O), each weighted by its heat-trapping capacity over a 100-year period), i.e., CO_2e per kilogram of animal product, or carbon footprint. In some cases, decreasing the emissions of CO_2e per kilogram of animal product has allowed lowering of the total number of animals sufficiently in a country or region so as to decrease the total emissions of CO_2e of the livestock industries e.g., Capper et al. [9]. However, in various other cases, the decrease in the emissions of CO_2e per unit of animal product occurring as a consequence of intensification was insufficient to compensate for the increase in animal production, resulting, therefore, in increased total CO_2e emissions from milk and beef production [17]. Whilst producing meat and milk with a lower carbon footprint is an important goal, intensification of animal production alone is unlikely to stop the increase in total emissions of GHG from ruminant production, much less mitigate them. Specific additional measures to ameliorate the emissions of CH_4 and other GHG from the livestock industry are also needed.

3. Mitigation of Enteric Methane Emissions

It is challenging to reconcile the objectives of decreasing total emissions of enteric CH_4 from ruminant production and at the same time increase the global supply of animal products. Therefore, several strategies to mitigate enteric CH_4 emissions from ruminants are being investigated: increasing feed efficiency, genetic selection of animals with lower CH_4 production, modifying diet formulation and concentrate and forage processing, grazing management, the addition of oils to the diet, use of chemical inhibitors of methanogenesis, dietary inclusion of algae with antimethanogenic compounds, alternative electron acceptors, phytocompounds, defaunation (elimination of rumen protozoa), immunization against methanogens, early-life interventions, and archaeal phages, among others. For more information, readers are referred to various excellent published reviews [18–25].

The effectiveness of all [26–28] or some [29–34] enteric CH_4 mitigation strategies currently available has been quantified through various meta-analyses. In their meta-analysis, Arndt et al. [27] identified increasing the individual animal feed intake (by, on average, 58%, without altering the composition of the diet) as the most effective strategy to decrease the CH_4 intensity of milk production (by, on average, 16.7%), while simultaneously increasing animal productivity. Secondly, they identified the utilization of inhibitors of methanogenesis (including 3-nitrooxypropanol (3-NOP) and bromoform-containing red algae *Asparagopsis* spp.) as the most effective strategy to decrease total daily emissions of CH_4 per animal (by, on average, 35.2%) and emissions of CH_4 per kilogram of milk produced (by, on average, 31.8%), without negatively affecting animal productivity [27].

Using the average decreases in CH_4 production found in their meta-analysis, Arndt et al. [27] estimated that the adoption of increased feed intake or inhibitors of methanogenesis, or both antimethanogenic measures in combination, could allow containing of global temperature increase by 1.5 °C by 2030 but not by 2050, even if applied under an unrealistic scenario of global 100% adoption [27]. The conclusions of the analysis by Arndt et al. [27] illustrate the challenges and difficulties of increasing ruminant production while decreasing the emissions of enteric CH_4 and CO_2e.

4. Projection of Global Enteric Methane Emission under Different Scenarios of Intensification and Adoption of Inhibitors of Methanogenesis

A projection of emission of enteric CH_4 from beef, lamb, and bovine milk production and the sum of the three products between present time and 2050 was herein conducted, combining scenarios of production intensification and adoption of feed additives inhibiting rumen methanogenesis. The evolution of total global emissions of enteric CH_4 was simulated considering future increases in global production of beef, lamb, and bovine milk and, depending on each scenario of production intensification, future decreases in CH_4 intensity of beef, lamb, and bovine milk production.

The analysis considered two different scenarios for intensification of production: (i) constant production efficiency, assuming constant 2016 levels of CH_4 emission intensities of beef, lamb, and bovine milk, as reported by Chang et al. [7] for the 2014–2018 quinquennial period and (ii) improved production efficiency, assuming decreasing CH_4 emission intensities of beef, lamb, and bovine milk at constant rates. Constant annual rates of decline in CH_4 intensity of beef, lamb, and milk production were calculated from changes in global CH_4 intensities reported for those industries by Chang et al. [7] between the 2000–2004 and 2014–2018 quinquennials.

Chang et al. [7] estimated CH_4 intensities using three different methods; an average of the values reported with the three estimation methods was considered for CH_4 intensities of global lamb and milk production. As for beef production, Chang et al. [7] reported an increase in CH_4 intensity between the 2000–2004 and 2014–2018 quinquennials with one of their methods of estimation and decreases with the other two, with increasing CH_4 intensity on average. Assuming that the global CH_4 intensity of beef production between 2010 and 2050 is likely to decrease as beef production increases [10,35], only the two estimations reporting declining CH_4 intensity of global beef production were averaged for the projection herein conducted. Annual constant rates of decline in CH_4 intensity were then calculated between 2002 (corresponding to the 2000–2004 quinquennial period) and 2016 (corresponding to the 2014–2018 quinquennial period; Table 1).

Table 1. Estimated rates of change in production and methane (CH_4) intensity of beef, lamb, and milk between 2010 and 2050.

Animal Product	Metric	Period Used for Estimation	Total Change (%)	Rate (%/yr)	Source Used for Estimation
Beef	Production	2010–2050	58	1.15	FAO [13]
Lamb	Production	2010–2050	78	1.45	FAO [13]
Milk	Production	2010–2050	58	1.15	FAO [13]
Beef	CH_4 intensity	2002–2016	−0.52	−0.037	Chang et al. [7] [1]
Lamb	CH_4 intensity	2002–2016	−7.27	−0.54	Chang et al. [7] [2]
Milk	CH_4 intensity	2002–2016	−9.55	−0.71	Chang et al. [7] [2]

[1] Average of the two negative estimates. [2] Average of three estimates.

Both scenarios of production efficiency (constant or declining CH_4 intensity) were combined with four different scenarios of adoption of feed additive inhibitors of methanogenesis as the most potent strategy available to mitigate total emissions of enteric CH_4 per animal [26,27]: 0, 25, 50, or 100% as the theoretical maximum worldwide adoption of feed additive inhibitors of methanogenesis in global production of beef, lamb, bovine milk, and all three combined. Percentages of global adoption refer to the percentage of global beef, lamb, and bovine milk produced with the use of inhibitors of methanogenesis, rather than to the percentage of total animals receiving the additives. The adoption of inhibitors of methanogenesis was simulated to take place gradually within a five-year period beginning in 2023 and as a linear function of time until reaching the maximum for each scenario of adoption.

The CH_4 intensity of beef, lamb, or bovine milk in each year and scenario was multiplied by the global production of the corresponding animal product to obtain the total CH_4 emissions associated to each product. Initially, the use of inhibitors of methanogenesis in milk production was modeled as causing a 31.8% decrease in CH_4 intensity, as reported by Arndt et al. [27]. Arndt et al. [27] did not report an effect of inhibitors of methanogenesis on CH_4 intensity of meat production. For meat production, a 26.6% decrease in CH_4 intensity of beef and lamb production, as the average of the range (−13.2 to −39.9%) for the effect of 3-NOP reported by Almeida et al. [26] for CH_4 intensity of milk and body mass gain, was initially considered for the analysis. A stronger effect of methanogenesis inhibitors in dairy than in beef animals agrees with a meta-analysis conducted for 3-NOP [29]. Calculations of total CH_4 emissions were conducted based on decreases in CH_4 intensity caused by inhibitors of methanogenesis rather than decreases in daily CH_4 emitted per animal, because total CH_4 emissions can be calculated from the forecasted increases in total output of animal products and changes in CH_4 intensity, as opposed to using decreases in daily CH_4 per animal for calculating total CH_4, which would require predicting future changes in animal numbers.

From projected increases in production of beef, lamb, and milk of 58, 78, and 58%, respectively, between 2010 and 2050 [13], constant yearly rates of increase of 1.15, 1.45, and 1.15% for beef, lamb, and milk production, respectively, were calculated (Table 1). Given that Chang et al. [7] reported CH_4 intensity as kg CH_4 per kg of animal protein (of beef, lamb, or bovine milk), the future global production of each animal product was also calculated as kilograms of protein according to FAO [36].

Figures 1–3 depict the predicted evolution between present time and 2050 of enteric CH_4 emissions for different scenarios in which 0, 25, 50, or 100% of beef, lamb, and milk, or the sum of all three products (Figure 4) is produced with the use of inhibitors of rumen methanogenesis. The evolution of enteric CH_4 production for each scenario of adoption of inhibitors of methanogenesis was simulated under constant 2016 levels of CH_4 intensity or under decreasing CH_4 intensities, according to the rates in Table 1. The upper and lower targets of a 24 and 47% decrease in enteric CH_4 emissions by 2050 relative to 2010 levels, as required to maintain the global temperature increase within 1.5 °C [15], are shown as reference. The database used for the simulations is available [37].

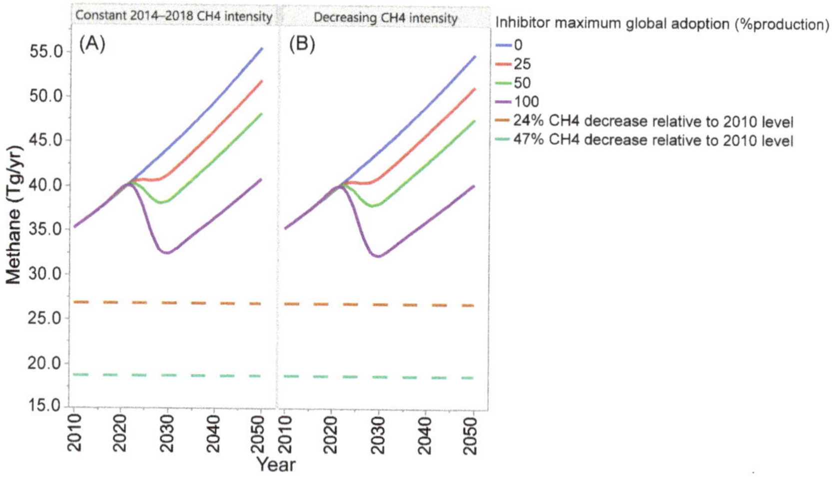

Figure 1. Predicted evolution of enteric methane (CH_4) emissions from global beef production between 2010 and 2050 considering (**A**) constant CH_4 intensity and (**B**) declining CH_4 intensity according to Table 1. Adoption of feed additive inhibitors of methanogenesis decreasing CH_4 intensity by 26.6% is simulated to occur at 0, 25, 50, or 100% of global beef production. Decreases in enteric CH_4 emissions of 24 and 47% relative to 2010 required to maintain a global temperature increase within 1.5 °C according to different socioeconomic scenarios and climate models [15] are shown in dashed lines.

Without introducing inhibitors of methanogenesis, the decrease in CH_4 intensity resulting from improving production efficiency can ameliorate CH_4 emissions from beef, lamb, bovine milk, and the sum of all three products by 0.70 (−1.3%), 1.58 (−17%), 7.80 (22%), and 10.1 (−10%) Tg/yr in 2050, compared to a scenario with constant CH_4 intensity. However, CH_4 emissions would still increase by 19.6 (+56%), 2.38 (+44%), 4.43 (+18.6), and 26.4 (+40.9%) Tg/yr from 2010 levels, in the same order, and would be 28.1 (+105%), 3.69 (+89%), 10.2 (+56%), and 41.9 (+85%) Tg/yr, in the same order, higher than the lower quartile CH_4 mitigation required to limit global temperature increase to 1.5 °C.

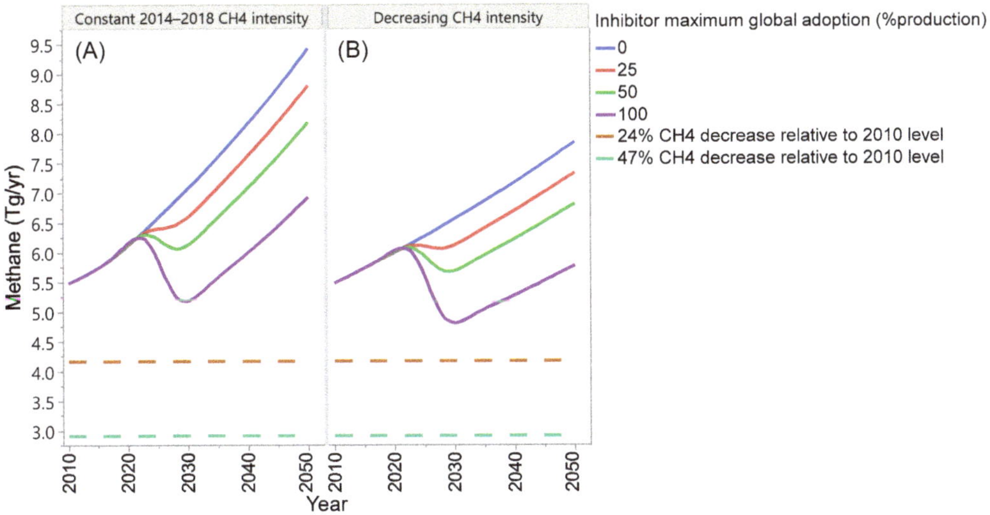

Figure 2. Predicted evolution of enteric methane (CH_4) emissions from global lamb production between 2010 and 2050 considering (**A**) constant CH_4 intensity and (**B**) declining CH_4 intensity according to Table 1. Adoption of feed additive inhibitors of methanogenesis decreasing CH_4 intensity by 26.6% is simulated to occur at 0, 25, 50, or 100% of global lamb production. Decreases in enteric CH_4 emissions of 24 and 47% relative to 2010 required to maintain a global temperature increase within 1.5 °C according to different socioeconomic scenarios and climate models [15] are shown in dashed lines.

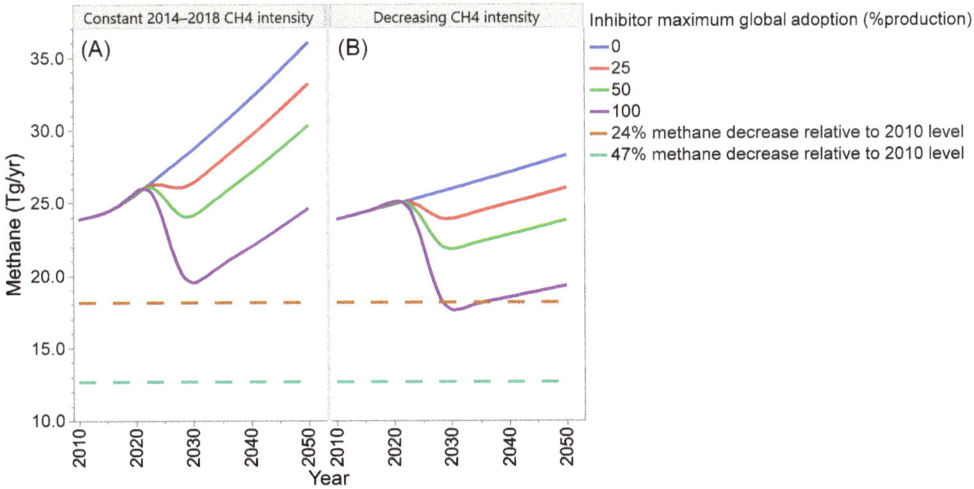

Figure 3. Predicted evolution of enteric methane (CH_4) emissions from global bovine milk production between 2010 and 2050 considering (**A**) constant CH_4 intensity and (**B**) declining CH_4 intensity according to Table 1. Adoption of feed additive inhibitors of methanogenesis decreasing CH_4 intensity by 31.8% is simulated to occur at 0, 25, 50, or 100% of global lamb production. Decreases in enteric CH_4 emissions of 24 and 47% relative to 2010 required to maintain a global temperature increase within 1.5 °C according to different socioeconomic scenarios and climate models [15] are shown in dashed lines.

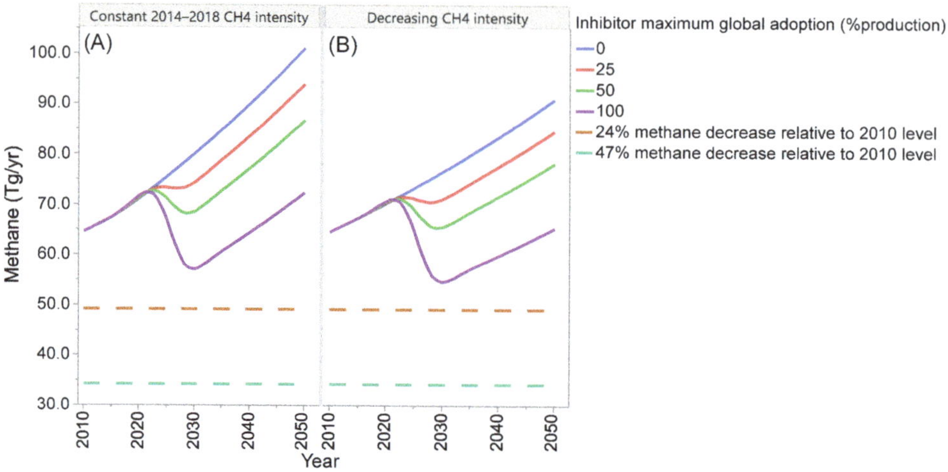

Figure 4. Predicted evolution of enteric methane (CH_4) emissions from global beef, lamb, and bovine milk production between 2010 and 2050 considering (**A**) constant CH_4 intensity and (**B**) declining CH_4 intensity according to Table 1. Adoption of feed additive inhibitors of methanogenesis decreasing CH_4 intensity by 26.6% for beef and lamb and by 31.8% for milk production is simulated to occur at 0, 25, 50, or 100% of global production. Decreases in enteric CH_4 emissions of 24 and 47% relative to 2010 required to maintain a global temperature increase within 1.5 °C according to different socioeconomic scenarios and climate models [15] are shown in dashed lines.

In this simulation, the mitigation of CH_4 emissions resulting from gains in production efficiency and resulting decreasing CH_4 intensity was estimated as minimal for beef production. The reason for this result is that the decline in CH_4 intensity between the 2000–2004 and 2014–2018 quinquennials for beef production estimated by Chang et al. (2021) [7] (Table 1), which was extrapolated to the 2023–2050 period in this simulation, was rather small. In fact, an increase in beef CH_4 intensity would have been considered had the calculation been conducted using the average of CH_4 intensity of the three estimations reported by Chang et al. (2021) [7] using three different methods. It is reasonable to think that the 2000–2004 to 2014–2018 tendency might change and greater gains in beef production efficiency and decreases in CH_4 intensity may result in the future; however, considering the results with lamb and bovine milk production, it appears likely that total CH_4 emissions from global beef production would still increase towards 2050, even under more optimistic improvements in CH_4 intensity. The adoption of practices that improve production efficiency is important to decrease the emissions of enteric CH_4 relative to a scenario with no improvements in CH_4 intensity; however, the emissions of enteric CH_4 would likely continue to grow if improvements in production efficiency are not accompanied by other measures.

That said, no projected scenario of enteric CH_4 emissions from beef or lamb production in this simulation would provide the required amelioration, even under an unrealistic 100% global adoption of inhibitors of methanogenesis (Figures 1 and 2). With 100% worldwide adoption of methanogenesis inhibitors, enteric CH_4 emissions from milk production would briefly decrease slightly below the upper 24% amelioration target around 2030 but would still not meet that target by 2050 (Figure 3). The same as beef and lamb production, total enteric CH_4 emissions from the sum of beef, lamb, and milk production was not projected to meet the minimum 24% decrease in enteric CH_4 emissions at any point in time between 2023 and 2050, even with a theoretical 100% worldwide adoption of inhibitors of methanogenesis (Figure 4).

5. Pronounced Inhibition of Rumen Methanogenesis with Feed Additives

Importantly, the use of feed additives inhibiting methanogenesis has allowed, in some studies, for considerably greater mitigation of enteric CH_4 production in comparison with the averages for methanogenesis inhibitors reported in the meta-analyses by Veneman et al. [28], Almeida et al. [26], and Arndt et al. [27]. Whilst, in the vast majority of studies evaluating inhibitors of methanogenesis, the extent of CH_4 decrease could be defined as moderate (i.e., ~30%), there have been various studies in which the inhibition of CH_4 production was considerably more profound (i.e., ~80% or more) [17]. Tables 2 and 3 summarize experiments in which the utilization of chemical inhibitors of methanogenesis or *Asparagopsis* spp. allowed decreasing CH_4 intensity of growth and fattening (Table 2) and milk production (Table 3) by 60% or more. There are studies not listed in Table 2 in which rumen methanogenesis was inhibited by 60% or more, but the animal performance was not reported and, thus, effects of methanogenesis inhibition on CH_4 intensity are not calculable (and, thus, cannot be used to calculate total emissions of CH_4 without knowing future changes in the number of animals); a more comprehensive list of experiments including treatments with inhibition of enteric CH_4 production equal to or greater than 60% is presented in Supplementary Table S1, demonstrating that pronounced inhibition of methanogenesis is possible. It is worth noting that, in 21 growth and fattening or maintenance experiments but in only two milk production experiments, a decrease of 60% or more in CH_4 production was obtained (Table S1), highlighting the need to investigate promoting pronounced inhibition of rumen methanogenesis in dairy cows.

Table 2. In vivo growth and fattening experimental treatments resulting in 60% or greater decrease in enteric methane (CH_4) production [1].

Reference	Animal, Diet	Inhibitor/Algae (g/kg Diet DM [2])	Experimental Period (d)	Inhibition Relative to Control Treatment (% Decrease in CH_4 Animal^{-1} d^{-1})	Inhibition Relative to Control Treatment (% Decrease in CH_4 kg ADG^{-1})	Performance		
						DMI	ADG	G:F
Trei et al. [38]	Lambs, mixed	2, 2, 2-trichloroacetamide (0.080)	90	67 [3]	67 [4]	NS [5]	NS	↑
Johnson et al. [39]	Steers, mixed	BCM (0.50)	28	~65 [6]	~68 [4]	NS	NS	-
Davies et al. [40]	Calves, mixed	ICI 13409 (0.20)	196	63 [3]	66 [4]	↓	↑	↑
Romero-Perez et al. [41]	Heifers, mixed	3-NOP (0.28)	112	59	60	R	NS	NS
Vyas et al. [42], finishing diet	Steers, high concentrate	3-NOP (0.2)	105	84	83	↓	↓	NS
Kinley et al. [43]	Steers, high concentrate	*Asparagopsis taxiformis* (1.8)	90	98	98	NS	↑	NS
Roque et al. [44]	Steers, high concentrate	*Asparagopsis taxiformis* (4.7)	63	82	83	↓	NS	NS
Alemu et al. [45]	Steers, high concentrate	3-NOP (0.108)	112	77	76 [4]	↓ [7]	↓ [7]	↑ [7]
Cristobal-Carballo et al. [46]	Calves, milk replacer, concentrate, partial mixed ration, pasture	Chloroform (0.050) plus 9, 10-anthraquinone (0.50)	84	90 [3]	90	NS	NS	-

[1] Only experiments reporting effects on performance and/or CH_4 intensity are presented. For a more complete list of experiments with at least one treatment with inhibition of CH_4 production equal to or greater than 60%, please refer to Table S1. [2] Abbreviations: 3-NOP = 3-nitrooxypropanol; ADG = average daily body mass gain; BCM = bromochloromethane; CH_4 = methane; DM = dry matter; DMI = dry matter intake; G:F = body mass gain per kilogram of dry matter intake; ICI 13409 = 2,4-bis(trichloromethy1)-benzo(1,3)dioxin-6-carboxylic acid. [3] Methane concentration, rather than production, was measured by rumenotomy [38,47] or air expelled in a hood [40]. [4] Calculated from reported results. [5] NS = effects reported as not significant ($p > 0.05$); ↑ = increase ($p < 0.05$); ↓ = decrease ($p < 0.05$); ↑ = tendency to increase ($0.05 \leq p < 0.10$); ↓ = tendency to decrease ($0.05 \leq p < 0.10$); - = not reported. [6] Daily average, estimated from results shown in a graph. [7] K. Beauchemin, pers. comm.

Modifying the previous analysis of prediction of enteric CH_4 emissions of Section 4, by considering an average efficacy of inhibitors of methanogenesis of 60% decrease in CH_4 intensity of beef, lamb, and milk production, the enteric CH_4 amelioration required to maintain global temperature increase within 1.5 °C by 2050 could be met only if inhibitors of methanogenesis were employed in nearly 85% of beef production (Figure 5), ~75% of lamb production (Figure 6), and ~60% of milk production (Figure 7). Reaching a 24% decrease in combined enteric CH_4 emissions from beef, lamb, and milk production by 2050 would require a worldwide adoption slightly greater than 75% of inhibitors of methanogenesis, decreasing CH_4 intensity by 60% (Figure 8). Evidently, these rates of adoption are highly unlikely. A more feasible scenario of 25% adoption of inhibitors of methanogenesis (e.g., in the most intensive production systems) with 60% efficacy would still decrease enteric CH_4 emissions of beef (−8.23 Tg/yr), lamb (−1.18 Tg/yr), and bovine milk (−4.24 Tg/yr) production by 15% relative to a scenario with no inhibitors of methanogenesis.

Table 3. In vivo milk production experimental treatments resulting in 60% or greater decrease in enteric methane (CH_4) production [1].

Reference	Animal, Diet	Inhibitor/Algae (g/kg Diet DM [2])	Experimental Period (d)	Inhibition Relative to Control Treatment (% Decrease in CH_4 Animal^{-1} d^{-1})	Inhibition Relative to Control Treatment (% Decrease in CH_4 kg FPCM^{-1})	Performance		
						DMI	MY	MY:F
Haisan et al. [48]	Cows, mixed	3-NOP (0.13)	28	60	61 [3]	NS [4]	NS	NS
Roque et al. [49]	Cows, mixed	*Asparagopsis armata* (10)	21	67	61 [3]	↓	↓	-

[1] Only experiments reporting effects on performance and/or CH_4 intensity are presented. For a more complete list of experiments with at least one treatment with inhibition of CH_4 production equal to or greater than 60%, please refer to Table S1. [2] Abbreviations: 3-NOP = 3-nitrooxypropanol; CH_4 = methane; DM = dry matter; DMI = dry matter intake; MY = milk yield; MY:F = milk yield per kilogram of dry matter intake. [3] Calculated from reported results. [4] NS = no significant ($p > 0.05$) effects; ↓ = decrease ($p < 0.05$); - = not reported.

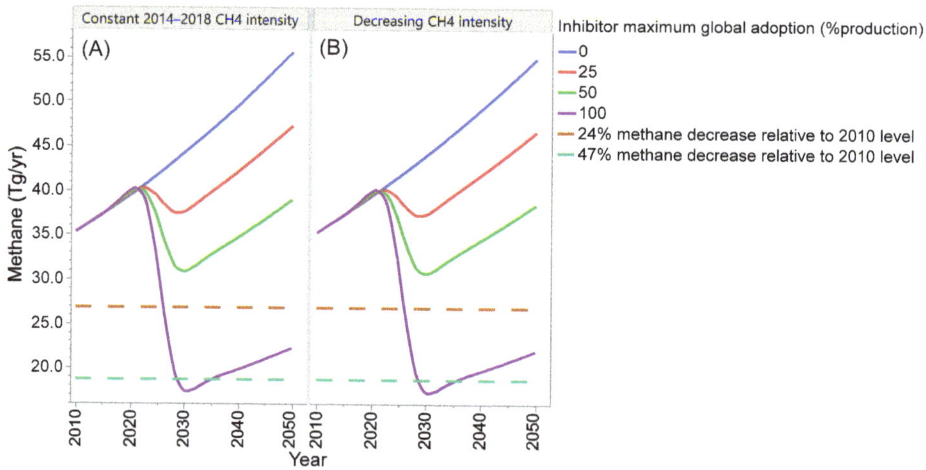

Figure 5. Predicted evolution of enteric methane (CH_4) emissions from global beef production between 2010 and 2050 considering (**A**) constant CH_4 intensity and (**B**) declining CH_4 intensity according to Table 1. Adoption of feed additive inhibitors of methanogenesis decreasing CH_4 intensity by 60% is simulated to occur at 0, 25, 50, or 100% of global beef production. Decreases in enteric CH_4 emissions of 24 and 47% relative to 2010 required to maintain a global temperature increase within 1.5 °C according to different socioeconomic scenarios and climate models [15] are shown in dashed lines.

It is thought that maximizing the use of inhibitors of methanogenesis can be an important component of a multifactorial effort to mitigate the emissions of enteric CH_4 from ruminant production systems. The rest of this document examines barriers to the adoption and enhanced efficiency of inhibitors of methanogenesis in ruminant production, possible solutions, and knowledge gaps and research hypotheses that can potentially generate knowledge to remove those barriers. Aspects pertaining to the adoption of inhibitors of methanogenesis that are discussed in the second part of this document are cost effectiveness and co-benefits (Section 6), delivery to animals and effectiveness in grazing systems (Section 7), and safety for animals, consumers, and the environment (Section 8). Research concerning the efficacy of inhibitors of methanogenesis is presented and discussed in Section 9.

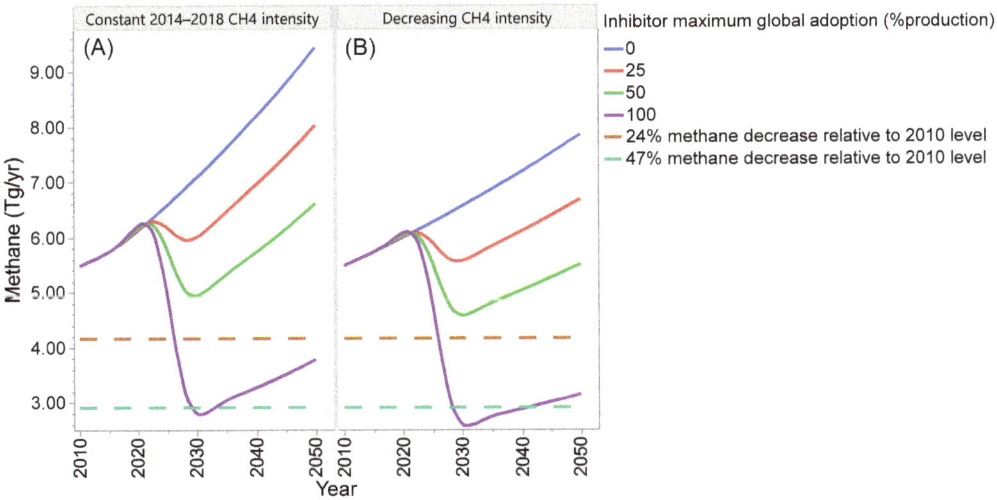

Figure 6. Predicted evolution of enteric methane (CH_4) emissions from global lamb production between 2010 and 2050 considering (**A**) constant CH_4 intensity and (**B**) declining CH_4 intensity according to Table 1. Adoption of feed additive inhibitors of methanogenesis decreasing CH_4 intensity by 60% is simulated to occur at 0, 25, 50, or 100% of global lamb production. Decreases in enteric CH_4 emissions of 24 and 47% relative to 2010 required to maintain a global temperature increase within 1.5 °C according to different socioeconomic scenarios and climate models [15] are shown in dashed lines.

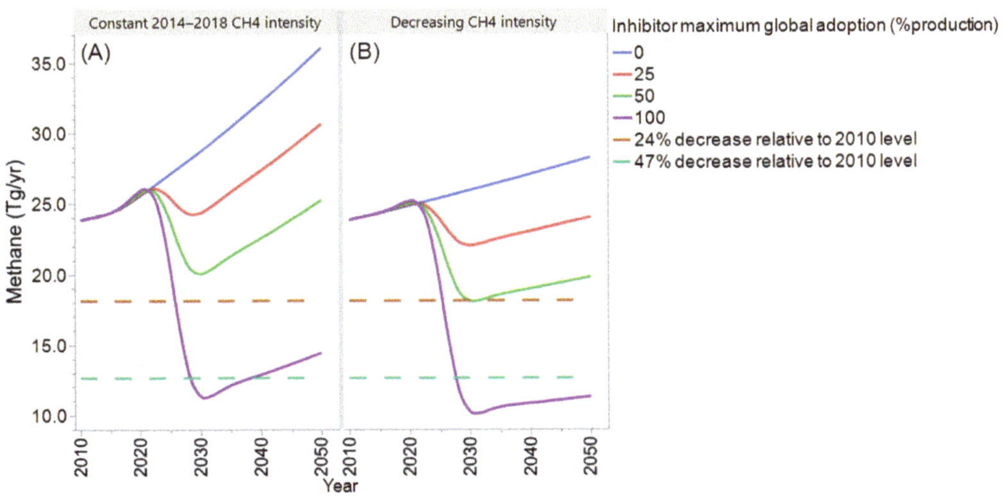

Figure 7. Predicted evolution of enteric methane (CH_4) emissions from global bovine milk production between 2010 and 2050 considering (**A**) constant CH_4 intensity and (**B**) declining CH_4 intensity according to Table 1. Adoption of feed additive inhibitors of methanogenesis decreasing CH_4 intensity by 60% is simulated to occur at 0, 25, 50, or 100% of global lamb production. Decreases in enteric CH_4 emissions of 24 and 47% relative to 2010 required to maintain a global temperature increase within 1.5 °C according to different socioeconomic scenarios and climate models [15] are shown in dashed lines.

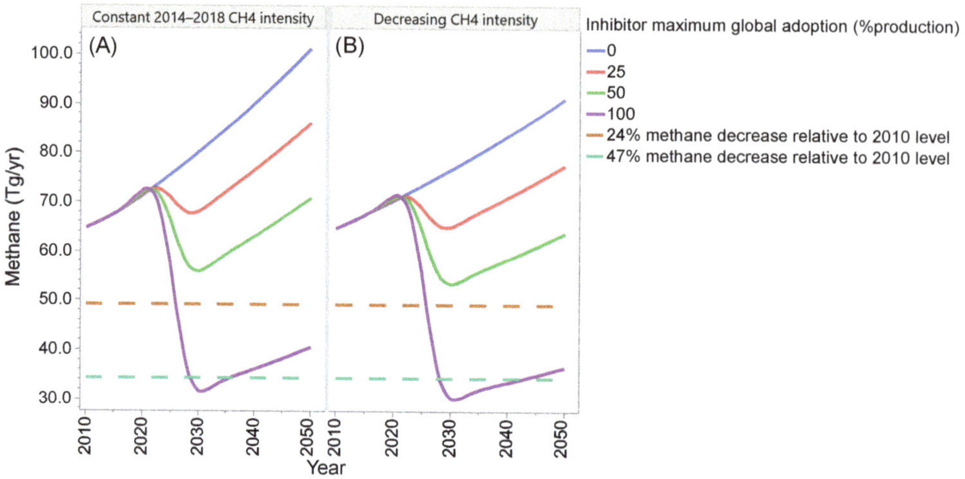

Figure 8. Predicted evolution of enteric methane (CH_4) emissions from global beef, lamb, and bovine milk production between 2010 and 2050 considering (**A**) constant CH_4 intensity and (**B**) declining CH_4 intensity according to Table 1. Adoption of feed additive inhibitors of methanogenesis decreasing CH_4 intensity by 60% for beef, lamb, and milk production is simulated to occur at 0, 25, 50, or 100% of global production. Decreases in enteric CH_4 emissions of 24 and 47% relative to 2010 required to maintain a global temperature increase within 1.5 °C according to different socioeconomic scenarios and climate models [15] are shown in dashed lines.

6. Cost Effectiveness and Co-Benefits of Inhibiting Methanogenesis

Successful adoption of antimethanogenic strategies requires that they are economically attractive to producers [50]. The inclusion of 3-NOP, *Asparagopsis*, or any other inhibitor of methanogenesis developed in the future in ruminant diets would raise feed costs, and if everything else remained unchanged, the use of antimethanogenic feed additives would be economically unattractive. Some possible means to overcome the added costs of antimethanogenic feed additives are:

1. Economic incentives;
2. Methanogenesis inhibition increasing feed efficiency;
3. Adjusting basal diet composition to the inhibition of methanogenesis.

6.1. Economic Incentives

It could be possible to overcome an increase in feed costs resulting from the inclusion of antimethanogenic feed additives in ruminant diets by establishing premium prices for meat and milk from animals fed inhibitors of methanogenesis and emitting less CH_4. Ultimately, consumers would be paying a higher price for environmentally friendly labeled meat or milk produced with lesser emissions of GHG. However, the relative size of these types of niche markets at a global scale is uncertain [51]. Markets for environmentally friendly labeled products were estimated as being of limited size in Europe [52]. Most of the growth in the production of animal products is expected to occur in the developing world [13], where niche markets for food products are generally marginal. The payment of a premium price for meat or milk products to encourage the adoption of inhibitors of methanogenesis may be feasible only in a relatively minor proportion of ruminant production markets situated in developed economies. Furthermore, higher prices of meat and milk associated with lower enteric CH_4 and CO_2e emissions might increase total CO_2e by further stimulating production in countries that already have a relatively high consumption of animal products.

Methane taxes are also a possibility to implement to stimulate the adoption of antimethanogenic measures, including feed additives inhibiting methanogenesis. A rising global tax has been proposed as an effective measure to mitigate emissions of CH_4. Models have predicted that responses in anthropogenic CH_4 abatement resulting from taxation would be largest in the fossil fuels sector, followed by agriculture and, lastly, the waste sector; in the ranking of modeled responses to CH_4 taxation by subsectors, enteric fermentation is second to coal [2,53]. The same as with the premium prices directed towards lowering enteric CH_4 emissions, it may be questionable whether enteric CH_4 taxation schemes would be implemented in developing economies increasing their production of ruminant products, at least in the short and medium term.

6.2. Methanogenesis Inhibition Increasing Feed Efficiency

Since at least the 1930s, ruminant nutritionists have known that CH_4 formation in the rumen is a loss of energy for the host animal and thus an inefficient process for ruminant production: "In the ruminant the waste products of digestion include the unused feed residues in the feces and relatively large amounts of methane gas, which is formed as a result of fermentation and serves no useful purpose" [54]. The realization of CH_4 formation in the rumen as an energy loss motivated in the 1960s and in following decades various studies seeking to inhibit methanogenesis in rumen fermentation to improve the efficiency of ruminant production. Blaxter and Czerkawski [55] proposed that "The almost saprophytic role which methanogenic organisms appear to play in the rumen, obtaining their energy from the end products of the microbial fermentation of carbohydrates and amino-acids, and producing waste products of methane and heat in large amounts, suggests very strongly that if their activity could be reduced without an impairment of cellulolysis, an increase in the productivity of ruminants could be obtained". It was not until the 1980s and 1990s, when the concern about livestock CH_4 emissions causing global warming [56,57] was raised, that the main goal of the research on the inhibition of rumen methanogenesis started to shift from improving energy efficiency to decreasing the environmental impacts of ruminant livestock e.g., Johnson and Johnson [58],McCrabb et al. [59].

Because ruminants lose between 2 and 12% of ingested gross energy as CH_4 [58], the possibility of enhancing feed efficiency through conserving energy that is otherwise lost as CH_4 could be a strong incentive to pay for the utilization of feed additives to inhibit rumen methanogenesis. However, examination of the evidence does not lead to conclusions about a consistent improvement in feed efficiency or animal performance when rumen methanogenesis is inhibited ([60]; Table 1, Table 2 and Table S1).

It has been suggested that the moderate inhibition of CH_4 production observed in most studies may not translate into large enough gains of net energy to elicit significant differences in feed efficiency or animal productivity [17,18]. Experiments with a much larger number of animals would thus be needed to detect significant differences in feed efficiency; in that regard, Alemu et al. [61], working with 4048 steers in a commercial feedlot, reported a tendency towards a 2.5% increase in feed efficiency when CH_4 production was moderately inhibited by 26%. It seems reasonable to think that energy savings resulting from moderate decreases in CH_4 production may not be detected in most experiments with a much smaller number of animals and lower statistical power. However, experiments with pronounced decreases in CH_4 production, still do not show consistent improvement in feed efficiency or animal performance (Table 1, Table 2 and Table S1).

Another factor that may contribute to explaining why inhibiting rumen methanogenesis does not consistently benefit animal productivity is that not all of the energy spared from CH_4 formation is incorporated into products nutritionally useful for the ruminant host animal. Most notably, inhibiting CH_4 production typically results in the release of dihydrogen (H_2) [62], which, in the typical rumen fermentation with functional methanogenesis, is a fermentation intermediate found at low concentration and the main electron donor for CH_4 production [63]. Accumulation of H_2 is potentially problematic because H_2 expelled is, in itself, a loss of energy. In 14 experiments in which the inhibition of CH_4

production was greater than 50%, energy lost as H_2 followed a nonlinear and variable relationship with energy lost in CH_4. Energy losses as H_2 could be described as moderate. For example, energy losses in H_2 at 80% inhibition of methanogenesis accounted for 11% [CI_{95} = 4.5, 17%] of the energy saved in CH_4 not formed. There has been speculation about incorporating H_2 into useful metabolic pathways through the use of electron acceptors or hydrogenotrophic microorganisms when methanogenesis is inhibited [64]. Benefits to animal productivity may not only depend on the sheer energy savings of H_2 incorporation, but also on the significance and metabolic fate of the absorbed sink of metabolic hydrogen for each type of animal, depending on its nutritional requirements [65].

Accumulation of H_2 resulting from inhibiting methanogenesis can impair the reoxidation of microbial NADH, increasing the NADH/NAD^+ ratio [66]. Insufficient availability of NAD^+ can theoretically halt fermentation [67]. Inhibiting methanogenesis in rumen microbial cultures induced H_2 accumulation and decreased fermentation as estimated from production of volatile fatty acids (VFA) [68]. In vivo, however, there is no conclusive evidence of negative effects of inhibiting methanogenesis on apparent digestibility or on VFA concentration adjusted by changes in DMI. It should be noted, however, that VFA concentration does not necessarily reflect VFA production. Changes in VFA production might be compensated by changes in rates of VFA absorption or passage, incorporation into microbial biomass, and changes in rumen volume; effects of inhibiting methanogenesis on the actual production of VFA have not been determined [17,60].

Apart from H_2, inhibition of methanogenesis in vitro [69] and in vivo e.g., Martinez-Fernandez et al. [70],Martinez-Fernandez et al. [71],Melgar et al. [72] has also resulted in increased concentrations of other electron carriers intermediate in rumen fermentation: formate, lactate, and ethanol. Lactate formed in propionate absorption through the rumen wall is used as a substrate for gluconeogenesis [73]. Direct absorption of lactate from the rumen seems to be influenced by adaptation to high concentrate diets [74]. Whether lactate could be absorbed from the rumen and utilized for gluconeogenesis in methanogenesis-inhibited animals is unknown. However, increases observed in lactate concentration in methanogenesis-inhibited rumens are much lower than what is observed in acidotic rumens and, if absorbed, it may likely have a relatively small influence on the ruminant's energy budget.

Results about the responses of formate and ethanol to methanogenesis inhibition are scarce. Formate can reach about 6–12 mM concentration in methanogenesis-inhibited rumens [70,71]; in other studies, it accumulated to a much lower concentration below 1 mM [72]. Ethanol concentration has been found to increase with methanogenesis inhibition, yet to relatively low levels [72,75]. It is unknown how the kinetics of formation and disappearance of formate and ethanol respond to pronounced methanogenesis inhibition and their metabolic fate, so as to establish the importance of these metabolites in the flow of carbon and metabolic hydrogen in the methanogenesis-inhibited fermentation and, fundamentally, their significance for the animal's energy budget.

6.3. *Adjusting Basal Diet Composition to the Inhibition of Methanogenesis*

Inhibiting methanogenesis is not an isolated intervention and causes profound changes in rumen fermentation. Alterations in both catabolic and anabolic processes may result in increased absorption of some nutrients and may open opportunities to decrease the need for them to be supplied by the basal diet.

Inhibiting rumen methanogenesis decreases the acetate-to-propionate concentration ratio, as predicted by the elevation of H_2 concentration [76] and confirmed in a meta-analysis of in vivo experiments [17]. Although propionate concentration adjusted by DMI did not respond to methanogenesis inhibition [17], it is still possible that, if increases in propionate production occur, they may have gone unnoticed when measuring propionate concentration because of compensatory changes occurring in propionate absorption. It is important to understand the responses of propionate production and absorption to methanogenesis inhibition because propionate is the main substrate for gluconeogenesis in ruminants [77]. If a positive response in propionate production to inhibiting methanogenesis could be shown experimentally through the use of labeled propionate, it would be important to understand the fate of the extra propionate absorbed and its metabolic consequences.

The meta-analysis by Loncke et al. [78] found that the formation of glucose from the sum of propionate, amino acids, and lactate increased at decreasing rates as their flow to the liver increased. This response suggests that glucogenic precursors may exceed the animal's demands for glucose as their availability increases. If inhibiting methanogenesis can be shown to increase propionate production in the rumen, perhaps basal diets could be modified to include less concentrates as glucogenic precursors and still match the animal requirements for glucose. This could allow decreasing feed costs in regions where concentrates are expensive. It might also prevent the decrease in DMI observed when CH_4 production is inhibited if the drop in DMI observed when inhibiting rumen methanogenesis [17,60] is caused by greater propionate oxidation in the liver acting as a satiety signal [79]. In regions of the world where cereal grains are not used for feeding ruminants, inhibiting methanogenesis might allow gluconeogenesis to be enhanced with forage-only diets.

Accumulated H_2 resulting from inhibiting rumen methanogenesis can also be incorporated into reductive acetogenesis, the reduction of CO_2 with H_2 to acetate. The addition of reductive acetogens to methanogenesis-inhibited in vitro fermentation was successful at incorporating accumulated H_2 into acetate formation [80–82]. Raju [83] showed the occurrence of reductive acetogenesis in sheep rumens and its increase when methanogenesis was inhibited by acetylene. Because of the higher H_2 threshold of reductive acetogens compared to the methanogens so far cultivated, it is expected that reductive acetogens could decrease H_2 accumulation, but H_2 concentration may still be higher compared to rumen fermentation with functional methanogenesis [64]. Animal production implications of enhancing reductive acetogenesis as a sink of metabolic hydrogen have been discussed [65].

The consequences of inhibiting rumen methanogenesis on microbial anabolism have received little attention. The incorporation of ammonium into the synthesis of microbial amino acids was stimulated by the methanogenesis inhibitor 9,10-anthraquinone in rumen cultures growing on starch but not on cellulose [84]. If this finding could be confirmed in vivo and with a broader range of inhibitors of methanogenesis and real diets fed to ruminants, it may be possible to replace greater proportions of expensive plant protein supplements with urea, again lowering feed costs and favoring cost effectiveness of the use of inhibitors of methanogenesis.

The effects of inhibiting methanogenesis on the rumen metabolism of fatty acids have potential implications for the quality of ruminant products. Decreases in milk fat percentage of vaccenic and rumenic acids, and mono- and polyunsaturated fatty acids observed when inhibiting rumen methanogenesis [72,85–89] suggest an increase in biohydrogenation, perhaps stimulated by the increased availability of reduced cofactors. This would be an undesirable consequence of the methanogenesis inhibition intervention, as the fatty acids profile in ruminant products would be richer in saturated fatty acids. Perhaps the decrease in mono- and polyunsaturated fatty acids could be lessened by adding to the diet sources rich in linolenic acid, such as fresh forages or linseed, when inhibiting rumen methanogenesis.

7. Adoption of Inhibitors of Methanogenesis in Grazing Systems

Globally, 37.4% of total enteric CH_4 emissions are generated by ruminants on free-ranging systems on rangelands and grasslands, 60.5% in mixed systems, and only 2.10% from beef cattle in feedlots [36]. Mixed systems can, in turn, comprise an ample range of production system typologies, from low-cost systems based on different proportions of pastures, concentrates, and agricultural and agro-industrial by-products to intensive dairy production operations with confined animals fed total mixed diets based on conserved forages and concentrates (as no dairy animals are classified in the Feedlot category in the

FAO GLEAM database [36], it is understood that dairy cows consuming mixed diets in intensive operations are classified in the Mixed systems category). Assuming that most intensive dairy operations using total mixed rations are located in North America and Europe, it can be estimated using data from the FAO GLEAM database [36] that about 7% of enteric CH_4 globally could be emitted from ruminants in intensive production systems with confined animals (i.e., feedlots and confined dairies; calculations not shown), the rest corresponding to extensive ranging systems and animals on pastures or crop residues, sometimes supplemented with concentrates or by-products.

For every peer-reviewed published study on enteric CH_4 mitigation conducted with grazing animals, 5.6 were conducted with confined animals (calculated from results by Vargas et al. [90]). Therefore, investment in research and development in the mitigation of enteric CH_4 production under intensive production conditions appears to be over-represented relatively to the contribution of confined production systems to global enteric CH_4 emissions and, conversely, information is lacking on enteric CH_4 mitigation in grazing systems. In particular, very little research has been conducted in extensive systems without any supplementation. This is especially important because, as discussed, antimethanogenic feed additives are the most potent means of decreasing enteric CH_4 emissions from ruminants [17,26,27,91] and they have been developed and evaluated to be delivered in feed supplemented to animals. Other means of delivery of antimethanogenic feed additives would have to be developed for extensively ranging animals without feed supplementation, such as salt and molasses lick blocks [92], boluses, or in drinking water [17]. Moreover, perhaps genes encoding for bromoform biosynthesis in *Asparagopsis* spp. could be genetically engineered in forages or directly in ruminants to deliver bromoform to the rumen in saliva or in rumen microbes. The latter possibilities, however, would require enough bromide content in the soil and in ingested forages, may have environmental implications, may affect the fitness and performance of bromoform-synthesizing plants, animals, or microbes, and could be both technically and economically difficult.

Other strategies to mitigate enteric CH_4 production in extensively ranging animals not receiving supplementation are being investigated, such as the selection of grazing animals producing less CH_4 [93], early life interventions with potential long-lasting effects [46,94], and immunization against rumen methanogens [95]. In general, mitigation of enteric CH_4 using these strategies has been mild or moderate, results have sometimes been contradictory, and more research is needed on their implications for animal productivity in different production systems. However, it is of much interest to continue conducting research in these antimethanogenic strategies which have the potential to be applied in extensive grazing systems to decrease the daily emissions of CH_4 per animal.

8. Safety and Other Aspects Important for the Adoption of Inhibitors of Methanogenesis

Adoption of feed additive inhibitors of methanogenesis towards pronounced mitigation of enteric CH_4 requires fulfilling various other aspects apart from effectiveness and persistency. Feed additives should not have negative effects on animal productivity and welfare, be safe for animals, consumers, and the environment, the decreases in enteric CH_4 emissions should not be compensated by upstream or downstream emissions of other GHG, additives must be possible to implement in the production systems being considered, must be approved by government agencies, be acceptable to consumers, and be economically attractive for producers to adopt.

3-Nitrooxypropanol is regarded as safe within recommended and experimentally evaluated doses [96,97]. Increasing the dose of 3-NOP would minimally increase upstream emissions of fossil fuel CO_2 associated with manufacturing and transporting 3-NOP, as, because of 3-NOP low levels of inclusion in diets, fossil fuel CO_2 associated with manufacturing and transporting 3-NOP is small in terms of CO_2e compared to CO_2e not emitted as CH_4 (calculations not shown).

Supplementing high doses of *Asparagopsis* can decrease DMI and milk production [49,89] and cause rumen mucosa abnormalities and inflammation [98,99]. Conversely, in other studies, supplementing *Asparagopsis* improved growth and feed efficiency [43,44]. Bromoform, the main CH_4-suppressing compound in *Asparagopsis*, is a suspected carcinogen and stratospheric-ozone-depleting agent [100], although the potential global ozone depletion caused by hypothetical global adoption of *Asparagopsis* was estimated to be relatively small [101]. Supplementing *Asparagopsis* has not resulted in the passage of bromoform to meat, milk, organs, or feces [43,44,49,89,98], with the exception of the first experimental day with non-adapted cows in the study by Muizelaar et al. [99]. However, supplementing *Asparagopsis* resulted in the passage of iodine and bromide to milk [89] and iodine to meat [44]. Bromoform is rapidly degraded by rumen cultures, mainly to dibromomethane [102], which is considered less toxic than bromoform [103]. An alternative might be the use of pure bromoform or dibromomethane (perhaps stabilized in a delivery complex, as it has been previously conducted with bromochloromethane [59]), which would allow the exact dosing of the active compound, independently of the content of bromoform in *Asparagopsis*. In addition, dosing pure bromoform (or dibromomethane) would avoid potential problems of excess iodine passing to milk and meat, although it would still result in bromide accumulation in milk.

9. Possibilities for Enhancing the Effectiveness of Inhibitors of Methanogenesis

Maximizing the efficacy of inhibitors of rumen methanogenesis, i.e., 60% or more, will be important to substantially mitigate enteric CH_4 emissions. Meta-analyses and individual studies show that the extent of inhibition of methanogenesis by 3-NOP [29,30,32] and *Asparagopsis* is positively related to their dose [43,44,49,89,98] and to the content of bromoform in the case of *Asparagopsis* [100]. Therefore, if high-end doses are used, the extent of inhibition of rumen methanogenesis by feed additives is potentially greater than the averages obtained in meta-analyses.

It must be noted, however, that responses to the dose of 3-NOP have not been linear in all individual studies evaluating multiple doses of 3-NOP [45,86,88,104] and it is possible that, under some conditions, the inhibition of methanogenesis may plateau at doses lower than maximal. It is important to understand the reasons, other than differing maximal doses examined in each study, behind the variation among studies in the linearity and the magnitude of the response of CH_4 decrease to the dose of the inhibitors. Aspects such as diet and animal, influencing the composition of the methanogenic community, may affect the magnitude of the responses to inhibitors of methanogenesis. For example, the response of CH_4 decrease to 3-NOP in beef and dairy animals has been shown to decline with increasing dietary NDF [29]. Conversely, using a very high dose of 1200 mg 3-NOP/kg of substrate DM in semicontinuous culture, Schilde et al. [105] decreased CH_4 production by 97% without observing interactions between the dose of 3-NOP and the percentage of concentrate in the substrate incubated; their results show that a high dose of 3-NOP overcame the expected lower inhibition with higher NDF.

Increasing the effectiveness of methanogenesis inhibitors can represent an avenue to decrease their cost, i.e., achieving more pronounced inhibition with current average doses or the same extent of inhibition with lower doses than current averages. Understanding the effects of 3-NOP on different methanogens [106,107], as well as elucidating the mechanisms that contribute to the resistance of methanogens to inhibitors [108], can help design means to improve their efficacy and cost effectiveness.

Differential sensitivity among different methanogens grown in pure culture to 3-NOP [109] and to the chemical inhibitor of methanogenesis 2-bromoethanesulfonate (BES) [83,110] has been reported before. Both 3-NOP [109] and BES [111] inhibit methanogenesis as structural analogs of methyl-coenzyme-M, a methylated cofactor involved in the last step of methanogenesis. Whilst inhibition caused by 3-NOP is persistent [41,42,86], inhibition caused by BES in sheep lasted for only 3 d, after which methane production returned to pretreatment levels [112].

Methanobrevibacter ruminantium M1, which has lost three genes required to synthesize coenzyme M [113] and, therefore, requires coenzyme M included in its growth medium [114], took up coenzyme M from the medium with a high-affinity transport system. Conversely, *Mbr. ruminantium* PS, which synthesizes coenzyme M, took up coenzyme M with a rate of less than 10% compared to *Mbr. ruminantium* M1 [115]. Mutants of *Methanococcus voltae* resistant to BES had a considerably reduced capacity to transport BES into the cell [116]. Inhibition of methane production by BES in *Mbr. ruminantium* M1 and *Methanosarcina* spp. could be diminished or reversed by the addition of coenzyme M to the medium, demonstrating a competition for transport between coenzyme M and BES [115,117]. The same as with BES, it appears conceivable that differences among methanogens in sensitivity to 3-NOP could also be related to the transport of 3-NOP into the cell and the ability of methanogen species to synthesize coenzyme M. It is possible that the effectiveness of 3-NOP at inhibiting methanogenesis is influenced, among other factors, by the proportion of coenzyme-M-synthesizing methanogens in the rumen methanogenic community and by the concentration of coenzyme M in rumen fluid.

In vivo work has also revealed shifts in the methanogenic community when inhibitors of methanogenesis are supplemented. Chloroform inhibited *Mbr. gottschalkii* more than *Mbr. ruminantium* [71], and both chloroform and 3-NOP were more inhibitory to hydrogenotrophic *Methanobrevibacter* spp. than to methylotrophic *Methanosphaera* spp. [75,106,118]. Those results agree with Duin et al. [109], who found that *M. ruminantium* was the most sensitive to 3-NOP of the methanogens they evaluated, *Msp. stadtmanae* was more resistant, and methylotrophic *Ms. barkeri* and hydrogenotrophic *Methanomicrobium mobile* were the most resistant. On the other hand, 3-NOP decreased the abundance of *Mmb. mobile* in rumen fluid [118]. Ungerfeld et al. [110] also found that methylotrophic *Ms. mazeii* was more resistant to BES and other inhibitors than *Mbr. ruminantium*, with *Mmb. mobile* being intermediate. It seems then that some chemical inhibitors evaluated may preferentially target hydrogenotrophic, over methylotrophic, methanogens. It is of much interest to understand if variation in the sensitivity of methanogens to particular chemical inhibitors is related to their metabolic pathway of CH_4 formation, i.e., hydrogenotrophic vs. methylotrophic methanogenesis, as the dietary content of methyl group precursors can influence the make-up of the methanogenic community and the relative importance of both methanogenic pathways. Furthermore, as H_2 thresholds differ among hydrogenotrophic, methyl-reducing, and methyl-fermenting methanogens [119], differential effects of 3-NOP on the different groups of methanogens could have implications for the extent of H_2 accumulation and release occurring as a consequence of inhibiting rumen methanogenesis.

Antimethanogenic compounds differ in their mechanisms of action. As discussed, 3-NOP [109] and BES [111] inhibit methyl-coenzyme M reductase through being structural analogs of methyl-coenzyme M, a cofactor present in all known methanogens. Methane halogenated analogs, such as chloroform or bromoform, react with cobamides and block the transfer of a methyl group from tetrahydromethanopterin to coenzyme M [120]. Derivatives of *p*-aminobenzoic acid inhibit the synthesis of methanogenic cofactor tetrahydromethanopterin [121], and statins inhibit methanogen growth by impairing membrane lipid synthesis [122]. Through the understanding of these mechanisms of action, it may be possible to design different combinations or rotations of antimethanogenic compounds to target specific methanogenic communities varying in composition depending on diet and animal. Because methanogens less inhibited by a particular compound may partially occupy the niche left by those methanogens inhibited more severely, it is conceivable that combining antimethanogenic compounds targeting different methanogens could result in synergic effects.

Apart from microbiological factors related to the sensitivity of different methanogens to inhibitors, the effectiveness of inhibitors of methanogenesis is also influenced by the daily pattern of inhibitor concentration in the rumen. Rumen concentration of inhibitors of methanogenesis is affected by the mode of administration, the time elapsed since the last feeding episode, the rates of feed ingestion and rumen fluid outflow, changes in rumen volume, and rates of metabolism and absorption of each specific compound. Almost all 3-NOP is metabolized to 1, 3-propanediol within 24 h and about 50% within 7 h [109]. Absorption of 3-NOP occurred in orally dosed rats, with plasma concentration peaking at

5–15 min after dosing [97], with no published results being available for ruminants. van Lingen et al. [123] modeled a peak in rumen concentration of 3-NOP of 0.055 mM 1.5 h after feeding, which gradually fell to 0 mM at about 12 h after feeding in animals fed twice per day a diet with 121 mg 3-NOP/kg DM. Bromoform was degraded in rumen microbial cultures by 70 and 90% after 30 min and 3 h incubation, respectively [102].

Predicted fluctuations in 3-NOP concentration in the rumen agree with the diurnal pattern of CH_4 emissions of animals fed once [88] or twice a day [45,61]. Dosing 3-NOP through the rumen cannula resulted in a relatively strong but short inhibition of methanogenesis, ultimately resulting in less than a 10% decrease in daily methane production, presumably because of the rapid washout of 3-NOP from the rumen [124]. Delivering methanogenesis inhibitors into the rumen in slow-release forms may result in more sustained concentration and greater effectiveness at inhibiting CH_4 production, even in animals fed total mixed rations. On the other hand, slow-release forms would likely increase manufacturing costs.

10. Final Remarks

Enteric CH_4 emissions from ruminants are a moving target for mitigation. This is because, as ruminant production increases, the decrease in total enteric CH_4 emissions necessary to contain global warming augments with time. A conclusion of simulations of projected enteric CH_4 emissions under constant or decreasing CH_4 intensity, and different extents of adoption and effectiveness of inhibitors of methanogenesis, was that antimethanogenic feed additives would have to be adopted at very large rates worldwide to decrease enteric CH_4 emissions as needed to limit the global temperature increase to 1.5 °C. In addition, antimethanogenic feed additives would have to consistently attain pronounced inhibition of rumen methanogenesis, which has been shown to be possible, although is beyond the average inhibition obtained in most studies. Inhibitors of methanogenesis are the single most potent enteric CH_4-mitigation strategy, but they will have to be complemented with continuous improvements in production efficiency and other CH_4 mitigation strategies. Carbon sequestration in soil can also be important to offset emissions of GHG, especially in degraded soils, but it should be taken into account that carbon accretion in soil does not proceed indefinitely. The side of the equation which has been assumed as fixed in this analysis, that is, the demand for ruminant products, should also be critically examined, bearing in mind the multiple nutritional, economic, social, and environmental implications of ruminant production. Reduction in the consumption of ruminant products may be considered in regions where it exceeds the nutritional requirements of human populations, as well as reductions in the use of arable land and human edible food in ruminant production and in food waste.

The use of antimethanogenic feed additives will increase feeding costs, which, everything else being the same, will discourage their adoption. This is especially true for extensive production systems that rely solely on grazing or on the use of low-cost by-products. For commercial extensive production systems, such as ranching, keeping low costs of production is key to remaining competitive in the business, whereas the possibilities for smallholders and pastoralists to increase production costs are limited by lack of finance and technology, access to markets, volatility of prices, and risks to sheer subsistence. Moreover, in intensive and semi-extensive operations, the adoption of antimethanogenic feed additives would likely be unattractive unless coupled with some benefit.

A premium price offered for meat and milk associated with lower CH_4 emissions could be attractive to producers so as to encourage the use of antimethanogenic feed additives. Niche markets ready to pay a higher price for environmentally produced livestock products are conceivable in developed economies but less likely in developing countries, where most of the growth in animal production is forecasted to take place. Methane taxes can be another economic incentive for CH_4 mitigation but, again, they may be more feasible to implement in developed economies, at least initially.

Research on understanding the alterations in the rumen and whole animal metabolism and physiology caused by inhibiting rumen methanogenesis may unveil new avenues for improving feed efficiency or reformulating basal diets and lowering their cost. This path, however, is relatively long-term, and its results are uncertain. Continued efforts in this direction are important but should also be accompanied by the implementation of proven antimethanogenic strategies causing moderate or mild decreases in emissions of enteric CH_4 and CO_2e, which can, at the same time, be productively and economically attractive.

There are important nutritional, economic, social, and environmental roles of ruminant production, its integration with crop production, and its ability to use nonarable land. The demand for ruminant meat and milk is increasing, especially in developing economies, thus also increasing the emissions of CH_4 from ruminant production. On the other hand, due to the relatively short half-life of CH_4 in the atmosphere, rapid action to decrease the emissions of enteric CH_4 from ruminant production offers an opportunity for short-term positive impact on climate change. It is necessary to investigate, develop, and adapt solutions, including antimethanogenic compounds, to the wide diversity of ruminant production systems worldwide.

Supplementary Materials: The following supporting information can be downloaded at: https://www.mdpi.com/article/10.3390/methane1040021/s1, Table S1: In vivo experimental treatments resulting in a 60% or greater decrease in enteric methane (CH_4) production.

Funding: This research was funded by Agencia Nacional de Investigación y Desarrollo ANID through project Fondecyt 1190574.

Institutional Review Board Statement: Not applicable.

Informed Consent Statement: Not applicable.

Data Availability Statement: Publicly available datasets were constructed by the author for this study. Data were made publicly available by the author on August 30, 2022, and can be found at: Open Science Framework: https://osf.io/drste/?view_only=2d0ee909617444a8b8568e50721d6e01, accessed on 30 August 2022.

Conflicts of Interest: The author declares no conflict of interest. The funder had no role in the design of the study; in the collection, analysis, or interpretation of data; in the writing of the manuscript; or in the decision to publish the results.

References

1. Hoegh-Guldberg, O.; Jacob, D.; Taylor, M.; Bindi, M.; Brown, S.; Camilloni, I.; Diedhiou, A.; Djalante, R.; Ebi, K.L.; Engelbrecht, F.; et al. Impacts of 1.5 °C global warming on natural and human systems. In *Global Warming of 1.5 °C*; Masson-Delmotte, V., Zhai, P., Pörtner, H.-O., Roberts, D., Skea, J., Shukla, P.R., Pirani, A., Moufouma-Okia, W., Péan, C., Pidcock, R., et al., Eds.; An IPCC Special Report on the Impacts of Global Warming of 1.5 °C above Pre-Industrial Levels and Related Global Greenhouse Gas Emission Pathways, in the Context of Strengthening the Global Response to the Threat of Climate Change, Sustainable Development, and Efforts to Eradicate Poverty; Intergovernmental Panel on Climate Change: Geneva, Switzerland, 2018; pp. 175–311.
2. United Nations Environment Programme and Climate and Clean Air Coalition. *Global Methane Assessment: Benefits and Costs of Mitigating Methane Emissions*; United Nations: Nairobi, Kenia, 2021; 173p.
3. IPCC. Summary for policymakers. In *Climate Change 2021: The Physical Science Basis. Contribution of Working Group I to the Sixth Assessment Report of the Intergovernmental Panel on Climate Change*; Masson-Delmotte, V., Zhai, P., Pirani, A., Connors, S.L., Péan, C., Berger, S., Caud, N., Chen, Y., Goldfarb, L., Gomis, M.I., et al., Eds.; Cambridge University Press: Cambridge, UK; New York, NY, USA, 2021; pp. 3–32. [CrossRef]
4. Szopa, S.; Naik, V.; Adhikary, B.; Artaxo, P.; Berntsen, T.; Collins, W.D.; Fuzzi, S.; Gallardo, L.; Kiendler-Scharr, A.; Klimont, Z.; et al. Short-lived climate forcers. In *Climate Change 2021: The Physical Science Basis. Contribution of Working Group I to the Sixth Assessment Report of the Intergovernmental Panel on Climate Change*; Masson-Delmotte, V., Zhai, P., Pirani, A., Connors, S.L., Péan, C., Berger, S., Caud, N., Chen, Y., Goldfarb, L., Gomis, M.I., et al., Eds.; Cambridge University Press: Cambridge, UK; New York, NY, USA, 2021; pp. 817–922. [CrossRef]
5. Saunois, M.; Stavert, A.R.; Poulter, B.; Bousquet, P.; Canadell, J.G.; Jackson, R.B.; Raymond, P.A.; Dlugokencky, E.J.; Houweling, S.; Patra, P.K.; et al. The global methane budget 2000–2017. *Earth Syst. Sci. Data* **2020**, *12*, 1561–1623. [CrossRef]
6. Reisinger, A.; Clark, H.; Cowie, A.L.; Emmet-Booth, J.; Gonzalez Fischer, C.; Herrero, M.; Howden, M.; Leahy, S. How necessary and feasible are reductions of methane emissions from livestock to support stringent temperature goals? *Philos. Trans. R. Soc. A* **2021**, *379*, 20200452. [CrossRef]

7. Chang, J.; Peng, S.; Yin, Y.; Ciais, P.; Havlik, P.; Herrero, M. The key role of production efficiency changes in livestock methane emission mitigation. *AGU Adv.* **2021**, *2*, e2021AV000391. [CrossRef]
8. Niu, M.; Kebreab, E.; Hristov, A.N.; Oh, J.; Arndt, C.; Bannink, A.; Bayat, A.R.; Brito, A.F.; Boland, T.; Casper, D.; et al. Prediction of enteric methane production, yield, and intensity in dairy cattle using an intercontinental database. *Glob. Chang. Biol.* **2018**, *24*, 3368–3389. [CrossRef] [PubMed]
9. Capper, J.L.; Cady, R.A.; Bauman, D.E. The environmental impact of dairy production: 1944 compared with 2007. *J. Anim. Sci.* **2009**, *87*, 2160–2167. [CrossRef]
10. Capper, J.L. The environmental impact of beef production in the United States: 1977 compared with 2007. *J. Anim. Sci.* **2011**, *89*, 4249–4261. [CrossRef]
11. Hristov, A.N.; Ott, T.; Tricarico, J.; Rotz, A.; Waghorn, G.; Adesogan, A.; Dijkstra, J.; Montes, F.; Oh, J.; Kebreab, E.; et al. Mitigation of methane and nitrous oxide emissions from animal operations: III. A review of animal management mitigation options. *J. Anim. Sci.* **2013**, *91*, 5095–5113. [CrossRef]
12. Gebbels, J.N.; Kragt, M.E.; Thomas, D.T.; Vercoe, P.E. Improving productivity reduces methane intensity but increases the net emissions of sheepmeat and wool enterprises. *Animal* **2022**, *16*, 100490. [CrossRef]
13. FAO. *World Livestock 2011. Livestock in Food Security*; Food and Agriculture Organization of the United Nations: Rome, Italy, 2011; 130p.
14. FAO. *The Future of Food and Agriculture—Alternative Pathways to 2050*; Food and Agriculture Organization of the United Nations: Rome, Italy, 2018; 224p.
15. IPCC. Summary for policymakers. In *Global Warming of 1.5 °C*; Masson-Delmotte, V., Zhai, P., Pörtner, H.-O., Roberts, D., Skea, J., Shukla, P.R., Pirani, A., Moufouma-Okia, W., Péan, C., Pidcock, R., et al., Eds.; An IPCC Special Report on the Impacts of Global Warming of 1.5 °C above Pre-Industrial Levels and Related Global Greenhouse Gas Emission Pathways, in the Context of Strengthening the Global Response to the Threat of Climate Change, Sustainable Development, and Efforts to Eradicate Poverty; Cambridge University Press: Cambridge, UK; New York, NY, USA, 2018; pp. 3–24. [CrossRef]
16. Jackson, R.B.; Saunois, M.; Bousquet, P.; Canadell, J.G.; Poulter, B.; Stavert, A.R.; Bergamaschi, P.; Niwa, Y.; Segers, A.; Tsuruta, A. Increasing anthropogenic methane emissions arise equally from agricultural and fossil fuel sources. *Environ. Res. Lett.* **2020**, *15*, 071002. [CrossRef]
17. Ungerfeld, E.M.; Beauchemin, K.A.; Muñoz, C. Current perspectives on achieving pronounced enteric methane mitigation from ruminant production. *Front. Anim. Sci.* **2022**, *2*, 795200. [CrossRef]
18. Beauchemin, K.A.; Ungerfeld, E.M.; Eckard, R.J.; Wang, M. Review: Fifty years of research on rumen methanogenesis: Lessons learned and future challenges for mitigation. *Animal* **2020**, *14*, s2–s16. [CrossRef] [PubMed]
19. Cottle, D.J.; Nolan, J.V.; Wiedemann, S.G. Ruminant enteric methane mitigation: A review. *Anim. Prod. Sci.* **2011**, *51*, 491–514. [CrossRef]
20. Clark, H. Nutritional and host effects on methanogenesis in the grazing ruminant. *Animal* **2012**, *7*, 41–48. [CrossRef] [PubMed]
21. Clark, H.; Eckard, R.J. Mitigating methane in a systems context. In Proceedings of the 4th Australasian Dairy Science Symposium, Christchurch, New Zealand, 31 August–2 September 2010; pp. 78–85.
22. Eckard, R.J.; Grainger, C.; De Klein, C.A.M. Options for the abatement of methane and nitrous oxide from ruminant production: A review. *Livest. Sci.* **2010**, *130*, 47–56. [CrossRef]
23. Goopy, J. Creating a low enteric methane emission ruminant: What is the evidence of success to the present and prospects for developing economies? *Anim. Prod. Sci.* **2019**, *59*, 1759–1776. [CrossRef]
24. Hristov, A.N.; Oh, J.; Firkins, J.L.; Dijkstra, J.; Kebreab, E.; Waghorn, G.; Makkar, H.P.S.; Adesogan, A.T.; Yang, W.; Lee, C.; et al. Mitigation of methane and nitrous oxide emissions from animal operations: I. A review of enteric methane mitigation options. *J. Anim. Sci.* **2013**, *91*, 5045–5069. [CrossRef]
25. Martin, C.; Morgavi, D.P.; Doreau, M. Methane mitigation in ruminants: From microbe to the farm scale. *Animal* **2010**, *4*, 351–365. [CrossRef]
26. Almeida, A.K.; Hegarty, R.S.; Cowie, A. Meta-analysis quantifying the potential of dietary additives and rumen modifiers for methane mitigation in ruminant production systems. *Anim. Nutr.* **2021**, *7*, 1219–1230. [CrossRef] [PubMed]
27. Arndt, C.; Hristov, A.N.; Price, W.J.; McClelland, S.C.; Pelaez, A.M.; Cueva, S.F.; Oh, J.; Dijkstra, J.; Bannink, A.; Bayat, A.R.; et al. Full adoption of the most effective strategies to mitigate methane emissions by ruminants can help meet the 1.5 °C target by 2030 but not 2050. *Proc. Natl. Acad. Sci. USA* **2022**, *119*, e2111294119. [CrossRef]
28. Veneman, J.B.; Saetnan, E.R.; Clare, A.J.; Newbold, C.J. MitiGate; an online meta-analysis database for quantification of mitigation strategies for enteric methane emissions. *Sci. Total Environ.* **2016**, *572*, 1166–1174. [CrossRef] [PubMed]
29. Dijkstra, J.; Bannink, A.; France, J.; Kebreab, E.; van Gastelen, S. Short communication: Antimethanogenic effects of 3-nitrooxypropanol depend on supplementation dose, dietary fiber content, and cattle type. *J. Dairy Sci.* **2018**, *101*, 9041–9047. [CrossRef] [PubMed]
30. Jayanegara, A.; Sarwono, K.A.; Kondo, M.; Matsui, H.; Ridla, M.; Laconi, E.B.; Nahrowi. Use of 3-nitrooxypropanol as feed additive for mitigating enteric methane emissions from ruminants: A meta-analysis. *Ital. J. Anim. Sci.* **2018**, *17*, 650–656. [CrossRef]
31. Lee, C.; Beauchemin, K.A. A review of feeding supplementary nitrate to ruminant animals: Nitrate toxicity, methane emissions, and production performance. *Can. J. Anim. Sci.* **2014**, *94*, 557–570. [CrossRef]

32. Kim, H.; Lee, H.G.; Baek, Y.C.; Lee, S.; Seo, J. The effects of dietary supplementation with 3-nitrooxypropanol on enteric methane emissions, rumen fermentation, and production performance in ruminants: A meta-analysis. *J. Anim. Sci. Technol.* **2020**, *62*, 31–42. [CrossRef]
33. Lean, I.J.; Golder, H.M.; Grant, T.M.D.; Moate, P.J. A meta-analysis of effects of dietary seaweed on beef and dairy cattle performance and methane yield. *PLoS ONE* **2021**, *16*, e0249053. [CrossRef]
34. Patra, A.K. Meta-analyses of effects of phytochemicals on digestibility and rumen fermentation characteristics associated with methanogenesis. *J. Sci. Food Agric.* **2010**, *90*, 2700–2708. [CrossRef]
35. Legesse, G.; Beauchemin, K.A.; Ominski, K.H.; McGeough, E.J.; Kroebel, R.; MacDonald, D.; Little, S.M.; McAllister, T.A. Greenhouse gas emissions of Canadian beef production in 1981 as compared with 2011. *Anim. Prod. Sci.* **2016**, *56*, 153–168. [CrossRef]
36. Global Livestock Environmental Assessment Model (GLEAM). Available online: www.fao.org/gleam/en/ (accessed on 22 July 2021).
37. Ungerfeld, E.M.; Prediction of Enteric Methane Emissions and Mitigation under Different Scenarios of Intensification and Adoption and Effectiveness of Inhibitors of Rumen Methanogenesis. Released on: 30 August 2022. Available online: https://osf.io/drste/?view_only=2d0ee909617444a8b8568e50721d6e01 (accessed on 30 August 2022).
38. Trei, J.E.; Parish, R.C.; Singh, Y.K. Effect of methane inhibitors on rumen metabolism and feedlot performance of sheep. *J. Dairy Sci.* **1971**, *54*, 536–540. [CrossRef]
39. Johnson, E.D.; Wood, A.S.; Stone, J.B.; Moran, E.T., Jr. Some effects of methane inhibition in ruminants (steers). *Can. J. Anim. Sci.* **1972**, *52*, 703–712. [CrossRef]
40. Davies, A.; Nwaonu, H.N.; Stanier, G.; Boyle, F.T. Properties of a novel series of inhibitors of rumen methanogenesis; in vitro and in vivo experiments including growth trials on 2,4-bis (trichloromethyl)-benzo(1,3)dioxin-6-carboxylic acid. *Brit. J. Nutr.* **1982**, *47*, 565–576. [CrossRef]
41. Romero-Perez, A.; Okine, E.K.; McGinn, S.M.; Guan, L.L.; Oba, M.; Duval, S.M.; Kindermann, M.; Beauchemin, K.A. Sustained reduction in methane production from long-term addition of 3-nitrooxypropanol to a beef cattle diet. *J. Anim. Sci.* **2015**, *93*, 1780–1791. [CrossRef] [PubMed]
42. Vyas, D.; McGinn, S.M.; Duval, S.M.; Kindermann, M.; Beauchemin, K.A. Effects of sustained reduction of enteric methane emissions with dietary supplementation of 3-nitrooxypropanol on growth performance of growing and finishing beef cattle. *J. Anim. Sci.* **2016**, *94*, 2024–2034. [CrossRef] [PubMed]
43. Kinley, R.D.; Martinez-Fernandez, G.; Matthews, M.K.; de Nys, R.; Magnusson, M.; Tomkins, N.W. Mitigating the carbon footprint and improving productivity of ruminant livestock agriculture using a red seaweed. *J. Clean. Prod.* **2020**, *259*, 120836. [CrossRef]
44. Roque, B.M.; Venegas, M.; Kinley, R.D.; de Nys, R.; Duarte, T.L.; Yang, X.; Kebreab, E. Red seaweed (*Asparagopsis taxiformis*) supplementation reduces enteric methane by over 80 percent in beef steers. *PLoS ONE* **2021**, *16*, e0247820. [CrossRef] [PubMed]
45. Alemu, A.W.; Shreck, A.L.; Booker, C.W.; McGinn, S.M.; Pekrul, L.K.D.; Kindermann, M.; Beauchemin, K.A. Use of 3-nitrooxypropanol in a commercial feedlot to decrease enteric methane emissions from cattle fed a corn-based finishing diet. *J. Anim. Sci.* **2021**, *99*, skaa394. [CrossRef]
46. Cristobal-Carballo, O.; McCoard, S.A.; Cookson, A.L.; Ganesh, S.; Lowe, K.; Laven, R.A.; Muetzel, S. Effect of methane inhibitors on ruminal microbiota during early life and its relationship with ruminal metabolism and growth in calves. *Front. Microbiol.* **2021**, *12*, 710914. [CrossRef]
47. Trei, J.E.; Scott, G.C.; Parish, R.C. Influence of methane inhibition on energetic efficiency of lambs. *J. Anim. Sci.* **1972**, *34*, 510–515. [CrossRef]
48. Haisan, J.; Sun, Y.; Guan, L.L.; Beauchemin, K.A.; Iwaasa, A.; Duval, S.; Barreda, D.R.; Oba, M. The effects of feeding 3-nitrooxypropanol on methane emissions and productivity of Holstein cows in mid lactation. *J. Dairy Sci.* **2014**, *97*, 3110–3119. [CrossRef]
49. Roque, B.M.; Salwen, J.K.; Kinley, R.; Kebreab, E. Inclusion of *Asparagopsis armata* in lactating dairy cows' diet reduces enteric methane emission by over 50 percent. *J. Clean. Prod.* **2019**, *234*, 132–138. [CrossRef]
50. Hristov, A.N.; Oh, J.; Lee, C.; Meinen, R.; Montes, F.; Ott, T.; Firkins, J.; Rotz, A.; Dell, C.; Adesogan, A.; et al. *Mitigation of Greenhouse Gas Emissions in Livestock Production—A Review of Technical Options for Non-CO_2 Emissions*; FAO: Rome, Italy, 2013; Volume 177, 231p.
51. Leahy, S.; Clark, H.; Reisinger, A. Challenges and prospects for agricultural greenhouse gas mitigation pathways consistent with the Paris agreement. *Front. Sustain. Food Syst.* **2020**, *4*, 69. [CrossRef]
52. Feucht, Y.; Zander, K. *Consumers' Attitudes on Carbon Footprint Labelling: Results of the SUSDIET Project*; Johann Heinrich von Thünen Institute, Federal Research Institute for Rural Areas, Forestry and Fisheries: Hamburg, Germany, 2017; 62p.
53. FAO. *Five Practical Actions towards Low-Carbon Livestock*; FAO: Rome, Italy, 2019.
54. Ritzman, E.G.; Benedict, F.G. *Nutritional Physiology of the Adult Ruminant*; Carnegie Institution of Washington: Washington, DC, USA, 1938; 200p.
55. Blaxter, K.L.; Czerkawski, J. Modification of the methane production of the sheep by supplementation of its diet. *J. Sci. Food Agric.* **1966**, *17*, 417–421. [CrossRef]
56. Hogan, K.; Hoffman, J.; Thompson, A. Methane on the greenhouse agenda. *Nature* **1991**, *354*, 181–182. [CrossRef]

57. Crutzen, P.J.; Aselmann, I.; Seiler, W. Methane production by domestic animals, wild ruminants, other herbivorous fauna, and humans. *Tellus B* **1986**, *38B*, 271–284. [CrossRef]
58. Johnson, K.A.; Johnson, D.E. Methane emissions from cattle. *J. Anim. Sci.* **1995**, *73*, 2483–2492. [CrossRef] [PubMed]
59. McCrabb, G.J.; Berger, K.T.; Magner, T.; May, C.; Hunter, R.A. Inhibiting methane production in Brahman cattle by dietary supplementation with a novel compound and the effects on growth. *Aust. J. Agr. Res.* **1997**, *48*, 323–329. [CrossRef]
60. Ungerfeld, E.M. Inhibition of rumen methanogenesis and ruminant productivity: A meta-analysis. *Front. Vet. Sci.* **2018**, *5*, 113. [CrossRef]
61. Alemu, A.W.; Pekrul, L.K.D.; Shreck, A.L.; Booker, C.W.; McGinn, S.M.; Kindermann, M.; Beauchemin, K.A. 3-Nitrooxypropanol decreased enteric methane production from growing beef cattle in a commercial feedlot: Implications for sustainable beef cattle production. *Front. Anim. Sci.* **2021**, *2*, 641590. [CrossRef]
62. Czerkawski, J.W. *An Introduction to Rumen Studies*, 1st ed.; Pergamon Press: Exeter, UK, 1986. [CrossRef]
63. Hungate, R.E. Hydrogen as an intermediate in the rumen fermentation. *Arch. Mikrobiol.* **1967**, *59*, 158–164. [CrossRef]
64. Ungerfeld, E.M. Metabolic hydrogen flows in rumen fermentation: Principles and possibilities of interventions. *Front. Microbiol.* **2020**, *11*, 589. [CrossRef]
65. Ungerfeld, E.M. A theoretical comparison between two ruminal electron sinks. *Front. Microbiol.* **2013**, *4*, 319. [CrossRef]
66. Hino, T.; Russell, J.B. Effect of reducing-equivalent disposal and NADH/NAD on deamination of amino acids by intact rumen microorganisms and their cell extracts. *Appl. Environ. Microbiol.* **1985**, *50*, 1368–1374. [CrossRef]
67. Nagaraja, T.G.; Newbold, C.J.; van Nevel, C.J.; Demeyer, D.I. Manipulation of ruminal fermentation. In *The Rumen Microbial Ecosystem*, 2nd ed.; Hobson, P.N., Stewart, C.S., Eds.; Blackie Academic & Professional: London, UK, 1997; pp. 524–632. [CrossRef]
68. Ungerfeld, E.M. Shifts in metabolic hydrogen sinks in the methanogenesis-inhibited ruminal fermentation: A meta-analysis. *Front. Microbiol.* **2015**, *6*, 37. [CrossRef] [PubMed]
69. Ungerfeld, E.M.; Rust, S.R.; Burnett, R. Use of some novel alternative electron sinks to inhibit ruminal methanogenesis. *Reprod. Nutr. Dev.* **2003**, *43*, 189–202. [CrossRef]
70. Martinez-Fernandez, G.; Denman, S.E.; Yang, C.; Cheung, J.; Mitsumori, M.; McSweeney, C.S. Methane inhibition alters the microbial community, hydrogen flow, and fermentation response in the rumen of cattle. *Front. Microbiol.* **2016**, *7*, 1122. [CrossRef]
71. Martinez-Fernandez, G.; Denman, S.E.; Cheung, J.; McSweeney, C.S. Phloroglucinol degradation in the rumen promotes the capture of excess hydrogen generated from methanogenesis inhibition. *Front. Microbiol.* **2017**, *8*, 1871. [CrossRef]
72. Melgar, A.; Harper, M.T.; Oh, J.; Giallongo, F.; Young, M.E.; Ott, T.L.; Duval, S.; Hristov, A.N. Effects of 3-nitrooxypropanol on rumen fermentation, lactational performance, and resumption of ovarian cyclicity in dairy cows. *J. Dairy Sci.* **2020**, *103*, 410–432. [CrossRef] [PubMed]
73. McDonald, P.; Edwards, R.A.; Greenhalgh, J.F.D.; Morgan, C.A.; Sinclair, L.A.; Wilkinson, R.G. *Animal Nutrition*, 7th ed.; Prentice Hall: Harlow, UK, 2010; 714p.
74. Qumar, M.; Khiaosa-Ard, R.; Pourazad, P.; Wetzels, S.U.; Klevenhusen, F.; Kandler, W.; Aschenbach, J.R.; Zebeli, Q. Evidence of in vivo absorption of lactate and modulation of short chain fatty acid absorption from the reticulorumen of non-lactating cattle fed high concentrate diets. *PLoS ONE* **2016**, *11*, e0164192. [CrossRef]
75. Martinez-Fernandez, G.; Duval, S.; Kindermann, M.; Schirra, H.J.; Denman, S.E.; McSweeney, C.S. 3-NOP vs. halogenated compound: Methane production, ruminal fermentation and microbial community response in forage fed cattle. *Front. Microbiol.* **2018**, *9*, 1582. [CrossRef]
76. Janssen, P.H. Influence of hydrogen on rumen methane formation and fermentation balances through microbial growth kinetics and fermentation thermodynamics. *Anim. Feed Sci. Technol.* **2010**, *160*, 1–22. [CrossRef]
77. Larsen, M.; Kristensen, N.B. Precursors for liver gluconeogenesis in periparturient dairy cows. *Animal* **2013**, *7*, 1640–1650. [CrossRef] [PubMed]
78. Loncke, C.; Nozière, P.; Vernet, J.; Lapierre, H.; Bahloul, L.; Al-Jammas, M.; Sauvant, D.; Ortigues-Marty, I. Net hepatic release of glucose from precursor supply in ruminants: A meta-analysis. *Animal* **2020**, *14*, 1422–1437. [CrossRef]
79. Allen, M.S.; Bradford, B.J.; Oba, M. Board-invited review: The hepatic oxidation theory of the control of feed intake and its application to ruminants. *J. Anim. Sci.* **2009**, *87*, 3317–3334. [CrossRef]
80. Nollet, L.; Demeyer, D.I.; Verstraete, W. Effect of 2-bromoethanesulfonic acid and *Peptostreptococcus productus* ATCC 35244 addition on stimulation of reductive acetogenesis in the ruminal ecosystem by selective inhibition of methanogenesis. *Appl. Environ. Microbiol.* **1997**, *63*, 194–200. [CrossRef] [PubMed]
81. Le Van, T.D.; Robinson, J.A.; Ralph, J.; Greening, R.C.; Smolenski, W.J.; Leedle, J.A.Z.; Schaefer, D.M. Assessment of reductive acetogenesis with indigenous ruminal bacterium populations and *Acetitomaculum ruminis*. *Appl. Environ. Microbiol.* **1998**, *64*, 3429–3436. [CrossRef]
82. Lopez, S.; McIntosh, F.M.; Wallace, R.J.; Newbold, C.J. Effect of adding acetogenic bacteria on methane production by mixed rumen microorganisms. *Anim. Feed Sci. Technol.* **1999**, *78*, 1–9. [CrossRef]
83. Raju, P. Homoacetogenesis as an Alternative Hydrogen Sink in the Rumen. Ph.D. Thesis, University of Massey, Palmerston North, New Zealand, 2016.
84. Ungerfeld, E.M.; Aedo, M.F.; Muñoz, C.; Urrutia, N.L.; Martínez, E.D.; Saldivia, M. Inhibiting methanogenesis stimulated de novo synthesis of microbial amino acids in mixed rumen batch cultures growing on starch but not on cellulose. *Microorganisms* **2020**, *8*, 799. [CrossRef]

85. Abecia, L.; Toral, P.G.; Martín-García, A.I.; Martínez, G.; Tomkins, N.W.; Molina-Alcaide, E.; Newbold, C.J.; Yañez-Ruiz, D.R. Effect of bromochloromethane on methane emission, rumen fermentation pattern, milk yield, and fatty acid profile in lactating dairy goats. *J. Dairy Sci.* **2012**, *95*, 2027–2036. [CrossRef]
86. Hristov, A.N.; Oh, J.; Giallongo, F.; Frederick, T.W.; Harper, M.T.; Weeks, H.L.; Branco, A.F.; Moate, P.J.; Deighton, M.H.; Williams, S.R.; et al. An inhibitor persistently decreased enteric methane emission from dairy cows with no negative effect on milk production. *Proc. Natl. Acad. Sci. USA* **2015**, *112*, 10663–10668. [CrossRef]
87. Melgar, A.; Lage, C.F.A.; Nedelkov, K.; Räisänen, S.E.; Stefenoni, H.; Fetter, M.E.; Chen, X.; Oh, J.; Duval, S.; Kindermann, M.; et al. Enteric methane emission, milk production, and composition of dairy cows fed 3-nitrooxypropanol. *J. Dairy Sci.* **2021**, *104*, 357–366. [CrossRef]
88. Melgar, A.; Welter, K.C.; Nedelkov, K.; Martins, C.M.M.R.; Harper, M.T.; Oh, J.; Räisänen, S.E.; Chen, X.; Cueva, S.F.; Duval, S.; et al. Dose-response effect of 3-nitrooxypropanol on enteric methane emissions in dairy cows. *J. Dairy Sci.* **2020**, *103*, 6145–6156. [CrossRef]
89. Stefenoni, H.A.; Räisänen, S.E.; Cueva, S.F.; Wasson, D.E.; Lage, C.F.A.; Melgar, A.; Fetter, M.E.; Smith, P.; Hennessy, M.; Vecchiarelli, B.; et al. Effects of the macroalga *Asparagopsis taxiformis* and oregano leaves on methane emission, rumen fermentation, and lactational performance of dairy cows. *J. Dairy Sci.* **2021**, *104*, 4157–4173. [CrossRef]
90. Vargas, J.; Ungerfeld, E.; Muñoz, C.; DiLorenzo, N. Feeding strategies to mitigate enteric methane emission from ruminants in grassland systems. *Animals* **2022**, *12*, 1132. [CrossRef] [PubMed]
91. Hegarty, R.S.; Passetti, R.A.C.; Dittmer, K.M.; Wang, Y.; Shelton, S.; Emmet-Booth, J.; Wollenberg, E.; McAllister, T.A.; Leahy, S.; Beauchemin, K.; et al. An Evaluation of Emerging Feed Additives to Reduce Methane Emissions from Livestock; Climate Change, Agriculture and Food Security (CCAFS) and the New Zealand Agricultural Greenhouse Gas Research Centre (NZAGRC) Initiative of the Global Research Alliance (GRA): 2021; 104p. Available online: https://cgspace.cgiar.org/handle/10568/116489 (accessed on 30 August 2022).
92. Callaghan, M.J.; Tomkins, N.W.; Hepworth, G.; Parker, A.J. The effect of molasses nitrate lick blocks on supplement intake, bodyweight, condition score, blood methaemoglobin concentration and herd scale methane emissions in *Bos indicus* cows grazing poor quality forage. *Anim. Prod. Sci.* **2021**, *61*, 445–458. [CrossRef]
93. Rowe, S.; Hickey, S.; Jonker, A.; Hess, M.; Janssen, P.; Johnson, T.; Bryson, B.; Knowler, K.; Pinares-Patino, C.; Bain, W.; et al. Selection for divergent methane yield in New Zealand sheep—A ten year perspective. In Proceedings of the Association for the Advancement of Animal Breeding and Genetics Conference, Armindale, NSW, Australia, 27 October–1 November 2019; Association for the Advancement of Animal Breeding and Genetics: Armindale, NSW, Australia, 2019; pp. 306–309.
94. Meale, S.J.; Popova, M.; Saro, C.; Martin, C.; Bernard, A.; Lagree, M.; Yáñez-Ruiz, D.R.; Boudra, H.; Duval, S.; Morgavi, D.P. Early life dietary intervention in dairy calves results in a long-term reduction in methane emissions. *Sci. Rep.* **2021**, *11*, 3003. [CrossRef] [PubMed]
95. Baca-González, V.; Asensio-Calavia, P.; González-Acosta, S.; Pérez de la Lastra, J.M.; Morales de la Nuez, A. Are vaccines the solution for methane emissions from ruminants? A systematic review. *Vaccines* **2020**, *8*, 460. [CrossRef] [PubMed]
96. Thiel, A.; Rumbeli, R.; Mair, P.; Yeman, H.; Beilstein, P. 3-NOP: ADME studies in rats and ruminating animals. *Food Chem. Toxicol.* **2019**, *125*, 528–539. [CrossRef]
97. Thiel, A.; Schoenmakers, A.C.M.; Verbaan, I.A.J.; Chenal, E.; Etheve, S.; Beilstein, P. 3-NOP: Mutagenicity and genotoxicity assessment. *Food Chem. Toxicol.* **2019**, *123*, 566–573. [CrossRef]
98. Li, X.; Norman, H.C.; Kinley, R.D.; Laurence, M.; Wilmot, M.; Bender, H.; de Nys, R.; Tomkins, N. *Asparagopsis taxiformis* decreases enteric methane production from sheep. *Anim. Prod. Sci.* **2016**, *58*, 681–688. [CrossRef]
99. Muizelaar, W.; Groot, M.; van Duinkerken, G.; Peters, R.; Dijkstra, J. Safety and transfer study: Transfer of bromoform present in *Asparagopsis taxiformis* to milk and urine of lactating dairy cows. *Foods* **2021**, *10*, 584. [CrossRef]
100. Glasson, C.R.K.; Kinley, R.D.; de Nys, R.; King, N.; Adams, S.L.; Packer, M.A.; Svenson, J.; Eason, C.T.; Magnusson, M. Benefits and risks of including the bromoform containing seaweed *Asparagopsis* in feed for the reduction of methane production from ruminants. *Algal Res.* **2022**, *64*, 102673. [CrossRef]
101. Jia, Y.; Quack, B.; Kinley, R.D.; Pisso, I.; Tegtmeier, S. Potential environmental impact of bromoform from *Asparagopsis* farming in Australia. *Atmos. Chem. Phys.* **2022**, *22*, 7631–7646. [CrossRef]
102. Romero, P.; Belanche, A.; Hueso, R.; Ramos-Morales, E.; Salwen, J.K.; Kebreab, E.; Yañez-Ruiz, D.R. In In vitro rumen microbial degradation of bromoform and the impact on rumen fermentation. In Proceedings of the 8th International Greenhouse Gas & Animal Agriculture Conference, Orlando, FL, USA, 5–10 June 2022; University of Florida: Orlando, FL, USA, 2022; p. 253.
103. PubChem Annotation Record for Dibromomethane. Available online: https://pubchem.ncbi.nlm.nih.gov/source/hsdb/1334 (accessed on 8 July 2022).
104. Haisan, J.; Sun, Y.; Guan, L.L.; Beauchemin, K.A.; Iwaasa, A.; Duval, S.; Kindermann, M.; Barreda, D.R.; Oba, M. The effects of feeding 3-nitrooxypropanol at two doses on milk production, rumen fermentation, plasma metabolites, nutrient digestibility, and methane emissions in lactating Holstein cows. *Anim. Prod. Sci.* **2017**, *57*, 282–289. [CrossRef]
105. Schilde, M.; von Soosten, D.; Hüther, L.; Kersten, S.; Meyer, U.; Zeyner, A.; Dänicke, S. Dose–response effects of 3-nitrooxypropanol combined with low- and high-concentrate feed proportions in the dairy cow ration on fermentation parameters in a rumen simulation technique. *Animals* **2021**, *11*, 1784. [CrossRef]

106. Pitta, D.; Indugu, N.; Melgar, A.; Hristov, A.; Challa, K.; Vecchiarelli, B.; Hennessy, M.; Narayan, K.; Duval, S.; Kindermann, M.; et al. The Effect of 3-nitrooxypropanol, a Potent Methane Inhibitor, on Ruminal Microbial Gene Expression Profiles in Dairy Cows. *Microbiome* **2022**, *10*, 146. [CrossRef] [PubMed]
107. Pitta, D.W.; Melgar, A.; Hristov, A.N.; Indugu, N.; Narayan, K.S.; Pappalardo, C.; Hennessy, M.L.; Vecchiarelli, B.; Kaplan-Shabtai, V.; Kindermann, M.; et al. Temporal changes in total and metabolically active ruminal methanogens in dairy cows supplemented with 3-nitrooxypropanol. *J. Dairy Sci.* **2021**, *104*, 8721–8735. [CrossRef] [PubMed]
108. Long, F.; Cheung, C.Y.; Whitman, W.B.; Cook, G.M.; Ronimus, R.S. Using genome comparisons of wild-type and resistant mutants of *Methanococcus maripaludis* to help understand mechanisms of resistance to methane inhibitors. *Access Microbiol.* **2021**, *3*, 000244. [CrossRef]
109. Duin, E.C.; Wagner, T.; Shima, S.; Prakash, D.; Cronin, B.; Yanez-Ruiz, D.R.; Duval, S.; Rumbeli, R.; Stemmler, R.T.; Thauer, R.K.; et al. Mode of action uncovered for the specific reduction of methane emissions from ruminants by the small molecule 3-nitrooxypropanol. *Proc. Natl. Acad. Sci. USA* **2016**, *113*, 6172–6177. [CrossRef]
110. Ungerfeld, E.M.; Rust, S.R.; Boone, D.R.; Liu, Y. Effects of several inhibitors on pure cultures of ruminal methanogens. *J. Appl. Microbiol.* **2004**, *97*, 520–526. [CrossRef]
111. Gunsalus, R.P.; Romesser, J.A.; Wolfe, R.S. Preparation of coenzyme M analogs and their activity in the methyl coenzyme M reductase system of *Methanobacterium thermoautotrophicum*. *Biochemistry* **1978**, *17*, 2374–2377. [CrossRef] [PubMed]
112. Van Nevel, C.J.; Demeyer, D.I. Control of rumen methanogenesis. *Environ. Monit. Assess.* **1996**, *42*, 73–97. [CrossRef]
113. Leahy, S.C.; Kelly, W.J.; Altermann, E.; Ronimus, R.S.; Yeoman, C.J.; Pacheco, D.M.; Li, D.; Kong, Z.; McTavish, S.; Sang, C.; et al. The genome sequence of the rumen methanogen *Methanobrevibacter ruminantium* reveals new possibilities for controlling ruminant methane emissions. *PLoS ONE* **2010**, *5*, e8926. [CrossRef]
114. Taylor, C.D.; McBride, B.C.; Wolfe, R.S.; Bryant, M.P. Coenzyme M, essential for growth of a rumen strain of *Methanobacterium ruminantium*. *J. Bacteriol.* **1974**, *120*, 974–975. [CrossRef]
115. Balch, W.E.; Wolfe, R.S. Transport of coenzyme M (2-mercaptoethanesulfonic acid) in *Methanobacterium ruminantium*. *J. Bacteriol.* **1979**, *137*, 264–273. [CrossRef]
116. Santoro, N.; Konisky, J. Characterization of bromoethanesulfonate-resistant mutants of *Methanococcus voltae*: Evidence of a coenzyme M transport system. *J. Bacteriol.* **1987**, *169*, 660–665. [CrossRef]
117. Smith, M.R. Reversal of 2-bromoethanesulfonate inhibition of methanogenesis in *Methanosarcina* sp. *J. Bacteriol.* **1983**, *156*, 516–523. [CrossRef]
118. Gruninger, R.J.; Zhang, X.M.; Smith, M.L.; Kung, L.; Vyas, D.; McGinn, S.M.; Kindermann, M.; Wang, M.; Tan, Z.L.; Beauchemin, K.A. Application of 3-nitrooxypropanol and canola oil to mitigate enteric methane emissions of beef cattle results in distinctly different effects on the rumen microbial community. *Anim. Microbiome* **2022**, *4*, 35. [CrossRef] [PubMed]
119. Feldewert, C.; Lang, K.; Brune, A. The hydrogen threshold of obligately methyl-reducing methanogens. *FEMS Microbiol. Lett.* **2020**, *367*, fnaa137. [CrossRef]
120. Wood, J.M.; Kennedy, F.S.; Wolfe, R.S. The reaction of multihalogenated hydrocarbons with free and bound reduced vitamin B_{12}. *Biochemistry* **1968**, *7*, 1707–1713. [CrossRef] [PubMed]
121. Dumitru, R.; Palencia, H.; Schroeder, S.D.; DeMontigny, B.A.; Takacs, J.M.; Rasche, M.E.; Miner, J.L.; Ragsdale, S.W. Targeting methanopterin biosynthesis to inhibit methanogenesis. *Appl. Environ. Microbiol.* **2003**, *69*, 7236–7241. [CrossRef] [PubMed]
122. Miller, T.L.; Wolin, M.J. Inhibition of growth of methane-producing bacteria of the ruminant forestomach by hydroxymethylglutaryl∼scoa reductase inhibitors. *J. Dairy Sci.* **2001**, *84*, 1445–1448. [CrossRef]
123. Van Lingen, H.J.; Fadel, J.G.; Yáñez-Ruiz, D.R.; Kindermann, M.; Kebreab, E. Inhibited methanogenesis in the rumen of cattle: Microbial metabolism in response to supplemental 3-nitrooxypropanol and nitrate. *Front. Microbiol.* **2021**, *12*, 705613. [CrossRef] [PubMed]
124. Reynolds, C.K.; Humphries, D.J.; Kirton, P.; Kindermann, M.; Duval, S.; Steinberg, W. Effects of 3-nitrooxypropanol on methane emission, digestion, and energy and nitrogen balance of lactating dairy cows. *J. Dairy Sci.* **2014**, *97*, 3777–3789. [CrossRef] [PubMed]

methane

MDPI

Review

Reducing Enteric Methanogenesis through Alternate Hydrogen Sinks in the Rumen

Prasanta Kumar Choudhury [1,2], Rajashree Jena [1,2], Sudhir Kumar Tomar [1] and Anil Kumar Puniya [1,*]

1 Dairy Microbiology Division, ICAR-National Dairy Research Institute, Karnal 132001, India
2 Department of Dairy Technology, School of Agricultural and Bioengineering, Centurion University of Technology and Management, Paralakhemundi 761211, India
* Correspondence: akpuniya@gmail.com

Abstract: Climate change and the urgent need to reduce greenhouse gas (GHG) emission from agriculture has resulted in significant pressure on the livestock industry for advanced practices that are environmentally more sustainable. Livestock is responsible for more than 15% of anthropogenic methane (CH_4) emission via enteric fermentation and improved strategies for mitigating enteric CH_4 production therefore represents a promising target to reduce the overall GHG contribution from agriculture. Ruminal CH_4 is produced by methanogenic archaea, combining CO_2 and hydrogen (H_2). Removal of H_2 is essential, as its accumulation inhibits many biological functions that are essential for maintaining a healthy rumen ecosystem. Although several other pathways occur in the rumen, including reductive acetogenesis, propionogenesis, nitrate, and sulfate reduction, methanogenesis seems to be the dominant pathway for H_2 removal. Global warming is not the only problem associated with the release of CH_4 from ruminants, but the released GHG also represent valuable metabolic energy that is lost to the animal and that needs to be replenished via its food. Therefore, reduction of enteric CH_4 emissions will benefit not only the environment but also be an important step toward the efficient production of high-quality animal-based protein. In recent decades, several approaches, relying on a diverse set of biological and chemical compounds, have been tested for their ability to inhibit rumen methanogenesis reliably and without negative effects for the ruminant animal. Although many of these strategies initially appeared to be promising, they turned out to be less sustainable on the industrial scale and when implemented over an extended period. The development of a long-term solution most likely has been hindered by our still incomplete understanding of microbial processes that are responsible for maintaining and dictating rumen function. Since manipulation of the overall structure of the rumen microbiome is still a significant challenge targeting key intermediates of rumen methanogenesis, such as H_2, and population that are responsible for maintaining the H_2 equilibrium in the rumen could be a more immediate approach. Addition of microorganisms capable of non-methanogenic H_2 sequestration or of reducing equivalents are potential avenues to divert molecular H_2 from methanogenesis and therefore for abate enteric CH_4. However, in order to achieve the best outcome, a detailed understanding of rumen microbiology is needed. Here we discuss some of the problems and benefits associated with alternate pathways, such as reductive acetogenesis, propionogenesis, and sulfate and nitrate reduction, which would allow us to bypass H_2 production and accumulation in the rumen.

Keywords: hydrogenotrophy; methanogenesis; propionogenesis; reductive acetogenesis; sulfate reduction

1. Introduction

The rumen harbors a highly diverse and complex mixture of microorganisms, including archaea (10^8-10^9/mL), bacteria ($10^{10}-10^{11}$/mL), ciliate protozoa (10^6/mL), and fungi (10^6/mL), which facilitate the degradation of complex plant carbohydrates into small molecules [1] and ultimately provide metabolites that can be used by the ruminant animal [2–5]. Livestock are mainly fed with agricultural crops, which via microbial activity are converted to metabolic intermediates (i.e., volatile fatty acids (VFAs), such as acetate, butyrate and propionate, and hydrogen (H_2) and gaseous end products such as carbon dioxide (CO_2) and methane (CH_4) [6]. Increased microbial H_2 production and its subsequent accumulation, which can be promoted by a high-starch diet, have several detrimental effects on the rumen ecosystem and that can be attributed to a decrease in rumen pH triggered by starch fermentation. These effects include the deactivation of specific biomass-degrading enzymes from some of the most efficient fiber degraders of the rumen microbiome but also system-level responses, such as the reduction of feed conversion within the rumen [7,8]. Methanogens, a group of microbes belonging to the phylogenetic group of the archaea, combine molecular H_2 with CO_2 to produce CH_4 during methanogenesis, enabling the removal of H_2 from the system [9,10]. Although this removal of H_2 is important for maintaining a healthy rumen ecosystem, from the viewpoint of nutrient expenditure methanogenesis is a costly process, accounting for a gross energy intake loss of 2–12% in ruminants [11–14]. Since the annual production of enteric CH_4 accounts for ~15% of total anthropogenic CH_4 emissions [11,15], with CH_4 having a global warming potential 23-fold higher than that of CO_2, there is also a real and severe environmental cost associated with the energy of the enteric CH_4 that is released into the atmosphere.

Strategies and factors for CH_4 abatement have been reviewed in the past [1,9,12,16–25] and many of the strategies used to mitigate CH_4 from ruminants involve the use of antibiotics, ionophores [26], halogenated CH_4 analogues [27–29], heavy metals [30], lipid-rich materials such as coconut oil [31–33], probiotics [27], bacteriocin [34], and numerous chemicals [35,36]. Immunization against methanogens [37,38], elimination of ciliate protozoa (defaunation) both in in vivo and in vitro [39] and addition of acetogenic bacteria to rumen fluid [40–42] in in vitro experiments have also been tested. Use of toxic chemicals and antibiotics as inhibitors, although considered an option in the past, are no longer accepted due to rising concerns regarding their impact on the environment, the animal, and potentially on the consumer of the animal products [43]. Interventions using phage therapy, altering methanogenic diversity and chemogenomic approaches [6] are some of the more recent technologies, but the extent to which these processes remove and eliminate the produced H_2 still remains to be investigated. Therefore, a critical step for a successful CH_4 reduction strategy may be one that uses natural processes within the rumen. One such approach relies on establishing a non-methanogenic sink for H_2 produced during fermentation. This review will focus on these H_2 elimination pathways.

2. Hydrogen: A Key Player in Rumen Fermentation

H_2 concentration plays a major role in the regulation of microbial fermentation in the rumen [44–47]. The partial H_2 pressure is a key regulator of H_2 metabolism and the fate of ruminal H_2 disposal with dissolved H_2 gas and H_2 ion determining the redox potential of the rumen liquor. The efficient elimination of H_2 enhances fermentation by reducing its inhibitory effect on microbial growth and microbial degradation of plant material [48,49]. Destiny of H_2 liberation is associated with favorable thermodynamic changes and an inverse correlation between Gibbs free energy (ΔG^0) and the minimum partial H_2-pressure that is required for a reaction to continue: a reaction is considered to be thermodynamically more competitive when its requirement of H_2 partial pressure is low [50]. Due to this central regulatory role in rumen fermentation, H_2 can be considered to be the currency of ruminal fermentation [51]. Removal of the major fraction of the rumen H_2 occurs via the methanogenic archaea to CH_4, during which four moles of H_2 are consumed and converted into one mole of CH_4, which is then released into the atmosphere though eructation. During hydrogenotrophic methanogenesis, methanogens use CO_2 as carbon source and terminal electron acceptor and H_2 as electron donor. Other non-methanogenic rumen microbes, using CO_2 and other electron acceptors such as sulfate, nitrate, and fumarate, compete with methanogens for H_2, but they play a less dominant role in the removal of H_2 from the rumen ecosystem [52,53]. Non-methanogenic bacteria that use H_2 as electron donor include acetogens that reduce CO_2 to form acetate by the Wood-Ljungdahl pathway [54], sulfate-reducing bacteria (SRB) that reduce sulfate to hydrogen sulfide [55], nitrate-reducing bacteria (NRB) that reduce nitrate (NO_3) to ammonia (NH_4) and fumarate-reducing bacteria that use H_2 to form succinate [56,57]. Succinate can subsequentially be decarboxylated to propionate, which is a valuable nutrient for the ruminant animal [58], either by the succinate producer itself or it can be transferred to succinate users as an intercellular electron carrier [45]. Figure 1 summarizes the microbial pathways for H_2 removal from the rumen.

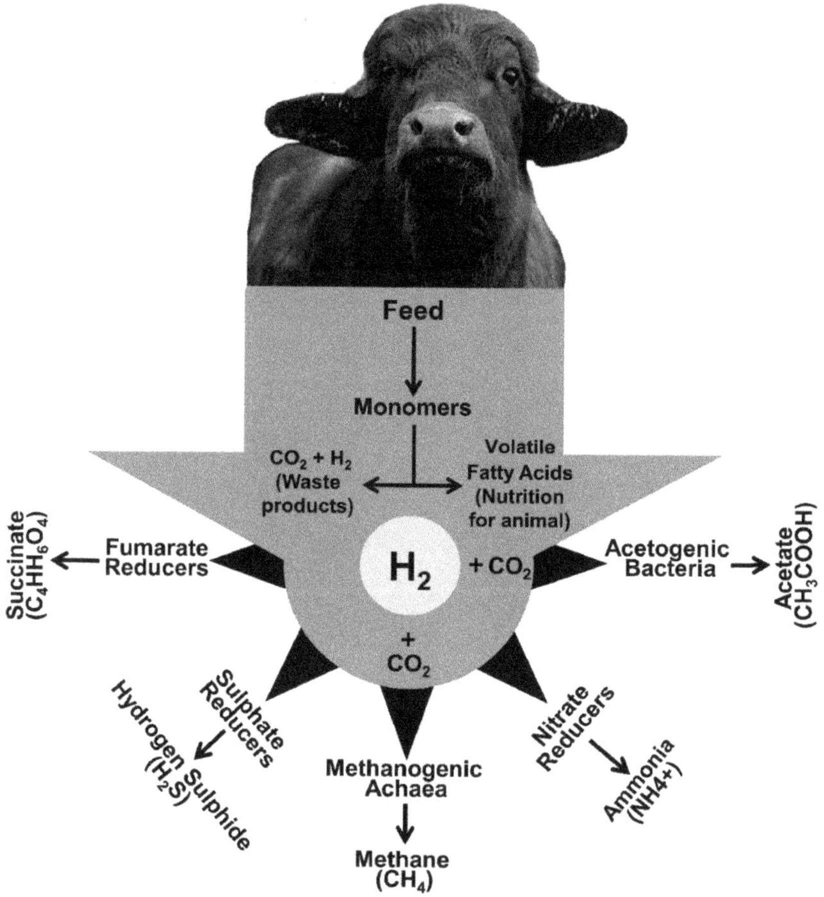

Figure 1. Major and minor H_2 and CO_2 sequestering pathways in rumen.

3. Alternative H_2 Sinks

The strategies to reduce CH_4 emission from enteric fermentation by non-methanogenic sinks are reviewed with respective mechanisms of action, thermodynamic changes; microorganism's involved, associated problems and anticipated management strategies are discussed (Table 1).

3.1. Reductive Acetogenesis

During reductive acetogenesis, also known as the Wood-Ljungdahl pathway or reductive acetyl-CoA pathway [7] H_2 and CO_2 are sequestered into acetate yielding energy for the ruminant host [19,54,59–63]. Due to the favorable energetics as well as absence of produced byproducts, reductive acetogenesis is a desirable way to eliminate excess H_2, and

H_2 concentration plays a vital role in deciding the fate of H_2 disposal. Acetogenesis can be autotrophic or heterotrophic, depending upon the type of substrate that is used. During autotrophic acetogenesis, two moles of CO_2 are reduced by four moles of H_2 to produce one mole of acetate ($4H_2 + 2CO_2 \rightarrow CH_3COOH + 2H_2O$) [60], whereas in heterotrophic acetogenesis, also referred to as homo-acetogenesis, one mole of hexose is converted to three moles of acetate, which is formed in a ratio of 2:1 from the oxidation of pyruvate and reduction of CO_2, respectively [64]. It is assumed that both autotrophic and heterotrophic acetogenesis occur simultaneously in the ruminant ecosystem. The Wood-Ljungdahl pathway has been described in diverse microbial ecosystems [59,65,66] where H_2 acts as an electron donor and CO_2 as an electron acceptor. The pathway contains two branches, methyl (western) and carbonyl (eastern) for synthesis of acetyl-CoA. The methyl branch is folate dependent where CO_2 reduced through formate and finally to methyltetrahydrofolate, while in carbonyl branch CO_2 reduced by carbon monoxide dehydrogenase to acetyl Co-A [62] (Figure 2). The change in Gibbs free energy during reductive acetogenesis is nearly −10.2 kJ/mol, while for methanogenesis from the same substrates is −68.3 kJ/mol [67]. This explains why reductive acetogenesis plays a minor role as hydrogenotrophic sink in the rumen when compared to microbial methanogenesis [68,69].

Table 1. H_2 sequestration sinks, involved mechanisms, associated problems and future directions.

Categories	Sub Groups	End Products	Microbes (Examples)	Overall Reaction	ΔG^0 (kJ)	Problems Associated	Management Strategies	Reference(s)
Methanogenic sinks	Methnogenesis	Methane (CH_4)	*Methanobrevibacter ruminantium*, *Methanomicrobium mobile*, *Methanobacterium bryantii*, *Methanobrevibacter smithii*, *Methanosarcina barkeri*, *Methanoculleus olentangyi*	$4H_2 + CO_2 \rightarrow CH_4 + 2H_2O$	−134.0	Source of ruminal CH_4, but not desirable as potent GHG.	Releases H_2 accumulation in rumen and need to be suppressed.	[2,4–6,11,12,18, 20,22–25,45,63]
Non-Methano-genic Sinks	Sulfate Reduction	Hydrogen Sulfide (H_2S)	*Desulfovibrio desulfuricans*, *D. vulgaris*, *Desulfatomaculum* spp.	$4H_2 + 2H^+ + SO_4 \rightarrow H_2S + 4\,H_2O$	−234.0	Undesirable reaction in rumen owing to toxicity of H_2S.	Most energy efficient sink in rumen dietary level and feeding strategy must taken into account.	[53,55,70–75]
	Reductive acetogens	Acetic acid (CH_3-COOH)	*Eubacterium limosum*, *Acetitomaculum ruminis*, *Blautia* spp, *Clostridium* spp., *Peptostreptococcus productus*, *Ruminococcus schinkii*, *Clostridium difficile*	$4H_2 + 2CO_2 \rightarrow CH_3COO^- + H^+ + 2H_2O$	−71.6	Desirable, but needs high levels of H_2 partial pressure.	Alteration of rumen microflora with a low H_2 threshold possessing capacity for reductive acetogenesis.	[7,10,19,47,54, 66,76,77]
	Nitrate Reduction	Ammonia (NH_4)	*Selenomonas ruminantium*, *Veillonella parvula* and *Wolinella succinogenes*	$4H_2 + 2H^+ + NO_3^- \rightarrow NH_4^+ + 3H_2O$	−519.0	Undesirable reaction in rumen owing to possible accumulation of toxic nitrite.	Gradual adaption of animal to supplement used and development of favorable microflora.	[53,67,70,78–91]
	Propionogenesis	Propionic acid (CH_3CH_2COOH)	*Fibrobacter succinogenes*, *Selenomonas ruminantium* ssp. *ruminantium*, *Selenomonas ruminantium* ssp. *lactilytica*, *Veillonella parvula* and *Wolinella succinogenes*	$C_6H_{12}O_6 + 2H_2 \rightarrow 2CH_3CH_2COOH + 2H_2O$	−84.0 (Fumarate to succinate)	Desirable reaction, but required substrate is costly.	Balancing minimum level in diet and dosing desired microbes governing propionate synthesis.	[57,92–103]

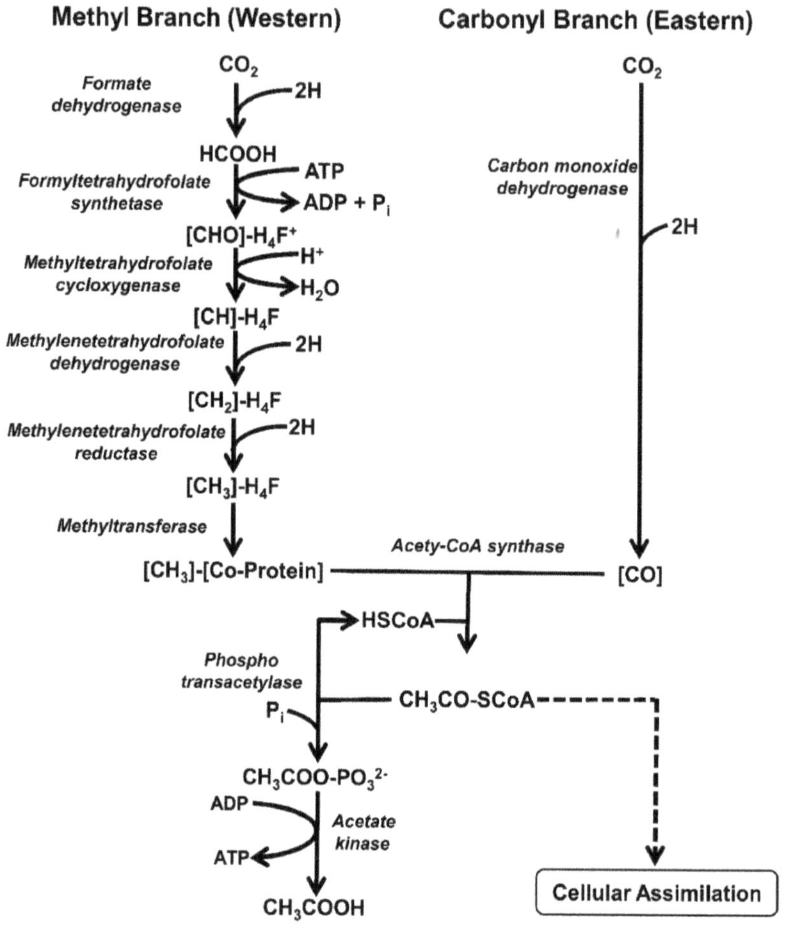

Figure 2. The Wood-Ljungdahl pathway of reductive acetogenesis.

Ruminal acetogens are not obligate in their substrate specificity and can contribute to H_2 production rather than H_2 consumption [41,104,105]. López et al. [106] reported that acetogenic bacteria can consume H_2 and CO_2 to form acetate significantly in the rumen when methanogenesis is inhibited. In the same study, they reported that an increase in the number of acetogenic bacteria cannot compete with methanogens. LeVan et al. [40] observed enhancement of in vitro reductive acetogenesis in incubations when methanogenesis was inhibited by BES with addition of rumen acetogen *Acetitomaculum ruminis* 190A4 and concluded that both selective inhibition of methanogenesis and addition of acetogens are crucial for the prevailing reductive acetogenesis under H_2 limiting conditions. However, Joblin [59] reported that ruminal acetogens dominate over methanogens and reduce CH_4 emission in vitro even if at low concentrations H_2. Competition for H_2 exists in rumen where acetogens are dominant hydrogenotrophs in the early rumen microbiota and methanogens replace them in later stage [61,76]. Fonty et al. [77] reported that reductive acetogens can maintain a functional rumen and replace methanogens as a sink for H_2 in methanogen free lambs and contributed to 21 to 25% to the rumen fermentation in vivo. However, Gagen et al. [76] observed in lambs that methanogen colonization does not significantly alter acetogen diversity isolated after 17 h after birth. Inhibition of ruminal methanogenesis and dosing of acetogens may lead significant increase in reductive acetogenesis [40,41]. Mitsumori et al. [107] reported a change in acetogen diversity in vivo in Holstein steers fed an antimethanogenic compound bromochloromethane (BCM). In another study, lambs removed from their mothers within 2 days of birth and kept in isolation appeared to use more metabolic H_2 via reductive acetogenesis and less CH_4 than conventionally raised lambs [108]. Other strategies may involve acetogen "enhancers" to provide acetogens with an advantage over methanogens, for example through the addition of yeast cells. *Saccharomyces cerevisiae* was reported to stimulate ruminal acetogens and their use of H_2 even in the presence of a methanogen in vitro [109]. Similarly, Yang et al. [110] observed enhanced acetogenesis and H_2 use increased the efficacy of acetogens in the presence of *S. cerevisiae* TWA4 strain.

In other gut environments, reductive acetogenesis is the major H_2 removal pathway and may be a useful source of potential acetogens to compete successfully with methanogens in the rumen [66,69,104,111,112] and thermodynamic control is not the single aspect for regulation of methanogen-acetogen interactions [69,104]. The prevailing H_2 gradient and the ability to grow mixotrophically in these environments may give acetogens a competitive advantage [67,113]. Acetogens isolated from eastern gray (*Macropus giganteus*), red kangaroos (*Macropus rufus*) and tammar wallaby (*Notamacropus eugenii*) have the capacity to compete with methanogens [111,112]. Methanogenesis was inhibited to an undetectable limit in reactors simulating the human gut by the addition of *Peptostreptococcus productus* [42]; interestingly this effect was diminished with ruminal fluid incubations [41]. A comparative analysis of acetogen isolated from ruminants (Ser5, Ser8, CA6 and SA11), marsupials (YE255, YE257 and YE266) and two reference isolates (*Acetitomaculum ruminis* and *Eubacterium limosum*) for H_2 use and acetate production showed that marsupial isolates (YE255, YE257 and YE266) are more efficient in using H_2; than ruminal isolates (CA6 and SA11) followed by reference isolates of acetogens [114]. Efficacy of acetogens to compete ruminal methanogenesis is observed to be source and strain dependent. Therefore, microbes with competitive ability at low H_2 partial pressure and/or addition from low CH_4 emitting animals must be taken into consideration for a fruitful reductive acetogenesis to establish in livestock.

3.2. Sulfur Reduction

Sulfate reduction is a thermodynamically highly favored process for H_2 removal in the rumen system [67]. Sulfate reducing bacteria (SRB) can be categorized based on the process they are employing for sulfate reduction into either assimilatory or dissimilatory SRB [115,116] (Figure 3). Both groups exist in rumen and facilitate the reduction of sulfur to hydrogen sulfite HSO_3^- and hydrogen sulfide H_2S. Dissimilatory reduction of sulfur compounds is used for energy generation, whereas during assimilatory sulfur reduction, sulfur compounds are incorporated into biological molecules that are necessary for cell survival [116,117]. In the formation of hydrogen sulfide, four moles of H_2 are consumed for each mole of H_2S generated. Energetically 1 ATP is consumed in the dissimilatory reduction of sulfate process to produce sulfide, whereas during the assimilatory process two ATP are used without generating H_2S. Dissimilatory reduction is the key route of sulfate metabolism in the rumen [118].

Dissimilatory SRB are strict anaerobic mesophiles, mostly Gram-negative, rod-shaped bacteria [119] that are ubiquitous in the digestive tract of mammals [55,120,121]. Members of the Desulfovibrionaceae (i.e., members of the genus *Desulfovibrio*) are the dominant SRB in ruminants [122,123]. Other abundant SRBs have been identified as belonging to the genus *Desulfotomaculum* and *Fusobacterium* [119,124]. Inclusion of SRB and/or sulfate in ruminant diet has shown to reduce CH_4 emission and enhance digestibility of the feed (Table 2). The recommended concentration of sulfur in growing beef cattle diet is 0.15% [125] and 0.14 to 0.26% in growing lamb diet [126]. Ruminant diets deficient in sulfur are connected with decreased microbial protein synthesis, digestibility, and lactate use [127,128]. Whanger and Matrone [129] reported microorganisms from sulfur-deficient animal contents could not synthesize butyrate and higher VFAs from acetate. Improved dry matter (DM) digestion, rumen fermentation and bacterial population in sheep fed a high sulfur diet were reported [130]. Patterson and Kung [131] observed adding sulfur at 0.3% DM improved cellulose digestion threefold in in vitro fermentations. Limited availability of sulfate in rumen has a direct effect on H_2 pressure in the rumen by suppressing microbial H_2 consumption and dietary increasing their concentration supports lowering CH_4 production. Change in microbial biomass and particularly an increase in SRB was observed with sulfate supplementation of the ruminant diet [70] and dissimilatory sulfate reduction was found to be proportional and also limited to the amount of sulfur available. Supplementation of sulfur to the regular diet fed to goats and lams resulted in the reduction of enteric CH_4 [70,71]. Paul et al. [124] reported supplementation of sulfate reducing bacteria SRBBR5, a strain capable of sulfate reduction, which resulted in the decrease CH_4 emissions from 2.66 to 1.64 mM CH_4/g DM truly digested after 72 h of fermentation without affecting methanogens and fungal population. In the same study, digestibility was reported to be increased significantly (15% in apparent digestibility and 40% in true digestibility), whereas H_2S concentration remained unaffected. Similarly, Wu et al. [72] reported that increased sulfur content of the animal diet resulted in decreased CH_4 emission (12.54 vs. 5.11 μM), total gas production ((39.1 vs. 27.1/mL culture), digestibility (63.0% vs. 51.5%)) and concentration of total VFAs in vitro, while increasing ammonia with no significant effect on archaea population.

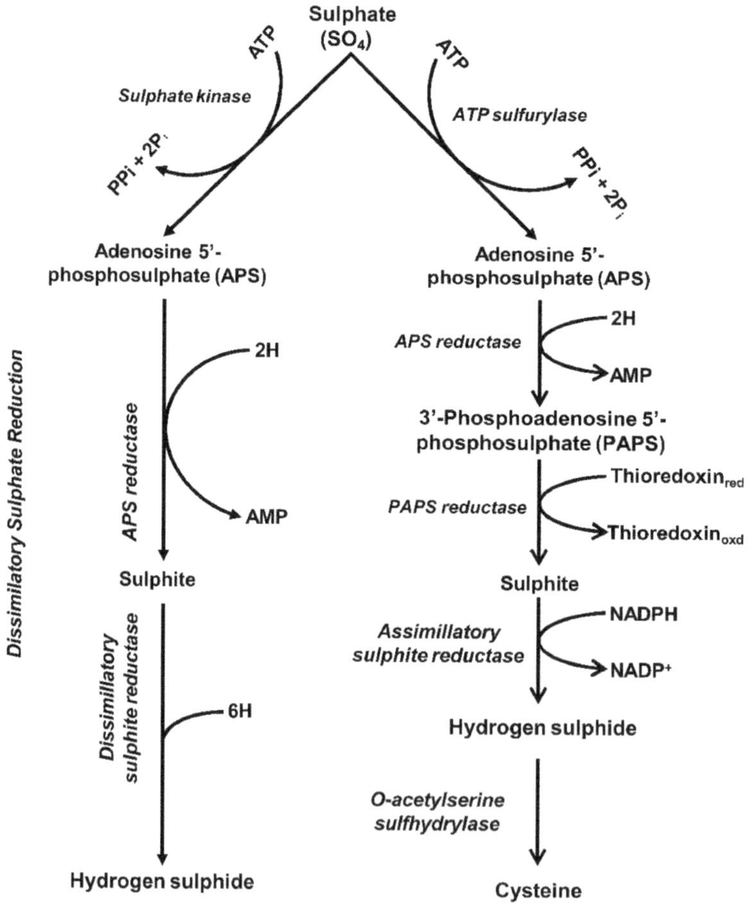

Figure 3. Dissimilatory and assimilatory route of sulfur reduction in rumen.

Table 2. Dietary sulfur, nitrate, fumarate and/or combinations on CH_4 production in in vitro or animal trials.

Dietary Supplements	Source and Level	Model	CH_4 Reduction (%)	References
Sulfur	Sulfate (2.6%)	Sheep	16	[70]
	Sodium sulfate (0.8%)	Goat	14.2	[71]
Nitrate	Pottasium nitrate (4%)	Sheep	23	[90]
	Pottasium nitrate (5%)	Cattle	43	[132]
	Pottasium nitrate (6%)	Cattle	27	[133]
	Nitrate (22 g/kg DM)	Cattle	32	[88]
	Nitrate (2.6%)	Sheep	32	[70]
	Calcium ammonium nitrate (2.84%)	Cattle	41	[134]
	Sodium nitrate (1.3 g/kg BW)	Sheep	50.4	[87]
	Nitrate (21 g/kg DM)	Cattle	16	[89]
	Calcium nitrate (3.8%/DM)	Goat	23.2	[71]
Fumarate	Fumaric acid (2% DM)	Cattle	23	[96]
	Encapsulated fumarate (10%)	Sheep	76	[135]
	Sodium fumarate (400 μM)	In vitro	17	[94]
	Sodium fumarate (500 μM)	In vitro	60	[136]
	Fumarate (3.5 g/L)	In vitro	38	[137]
	Sodium fumarate (6.2 mM)	In vitro	17	[95]
	Fumaric acid (8% DM)	Sheep	12	[138]
	Sodium acrylate	In vitro	8	[94]
	Sodium fumarate	In vitro	17	[94]
	Fumarate (10 mM)	In vitro	17	[139]
	Fumarate (30 mM)	In vitro	11	[140]
Combinations	Sulfur (2.6%) + Nitrate (2.6%)	Sheep	47	[70]
	Sodium sulfate (0.8%) + Calcium nitrate (3.8%)	Goat	34.9	[71]
	Sodium nitrate (1.3 g/kg BW) + GOS	Sheep	52.9	[87]
	Sodium nitrate (1.3 g/kg BW) + Nisin (3 mg/kg BW)	Sheep	56.3	[87]
	Sodium nitrate (5%) + Sulfur (0.4%)	Sheep	19.6	[141]
	Sodium nitrate (5%) + Sulfur (0.4%)	Goat	18.2	[141]

However, application of sulfur to the ruminant diet has some serious problems, especially when performed under not closely monitored conditions, since sulfur concentrations above a critical dose result in the H_2S, which is toxic to the host animal [53,72–74,142]. H_2S has limited solubility and is readily absorbed through the rumen wall into the blood stream [143], and therefore interrupts animal performance. The appearance of polioencephalomalacia (PEM) or cerebro-cortical necrosis occurs when sulfide travels through the blood to the brain, leading to death and contributes substantial economic loss to livestock industry [142]. Other associated problems of increased H_2S concentrations include adverse effect on the activity of respiratory enzymes (e.g., catalases, peroxidases, carbonic anhydrase, dopa-oxidases, dehydrogenases, and dipeptidases, cytochrome-c oxidase), production of sulfhemoglobin, depressed rumen motility, decreased mineral use, and several adverse effects on oxidative metabolism and energy generation in animals [115,144]. There have been reports of approaches that address and solve these H_2S toxicity problems. For example, feedlot cattle responded well to diets high in sulfate ferric citrate decrease [145] and passive immunization targeting SRB [146] adverse H_2S toxicity. However, significantly more work in this area is needed and efficient strategies to overcome H_2S toxicity still need to be identified before sulfur/sulfate can be considered to be a viable feed additive to inhibit methanogenesis.

3.3. Nitrate Reduction

The mechanism of nitrate (NO_3) reduction (Figure 4) can serve as an alternate pathway for lowering CH_4 emission due to NO_3 with a higher affinity for H_2 than CO_2 [78,147]. In anaerobic systems, nitrate reduction occurs by three distinct mechanisms [147], dissimilatory nitrate reduction to nitrogen gas (denitrification), assimilatory nitrate reduction (respiratory nitrate reduction, ANR) to ammonia and dissimilatory nitrate reduction to ammonia (Figure 4). Denitrification proceeds in a stepwise manner, in which nitrate (NO_3) is reduced to nitrite (NO_2), which then is reduced to nitric oxide (NO), which is further reduced to nitrous oxide (NO_2), and eventually to nitrogen gas (N_2). Although denitrification does not play a major role in the rumen under normal physiological conditions, trace amounts of nitrogen oxide can be measured when nitrate is abundant. In assimilatory nitrate reduction, the product of the enzymatic reaction remains in the organism itself to enable microbial protein synthesis. Nitrate is reduced to nitrite by NADH reduction reactions and nitrite is reduced to ammonia by respiratory ammonification, coupled with ATP production [79] to fulfill the energy needs associated with this form of nitrate reduction. In contrast to the energy producing dissimilatory nitrate reduction, high ammonia concentrations have an inhibitory effect on ANR. Hence ANR plays no major function on the rumen environment, where ammonia is abundant and rumen microorganisms will use this pathway to primarily synthesize sufficient ammonia to meet their requirements for biosynthesis and storage [148]. In this category the microorganism *Wolinella succinogenes* is the most comprehensively studied organism that carries out respiratory nitrite ammonification [79]. The genome sequences of *Wolinella succinogenes* showed that it is a close relative of the pathogenic epsilon proteobacteria *Helicobacter pylori* and *Campylobacter jejuni* and the first non-pathogenic bacteria whose genome sequence was determined [149].

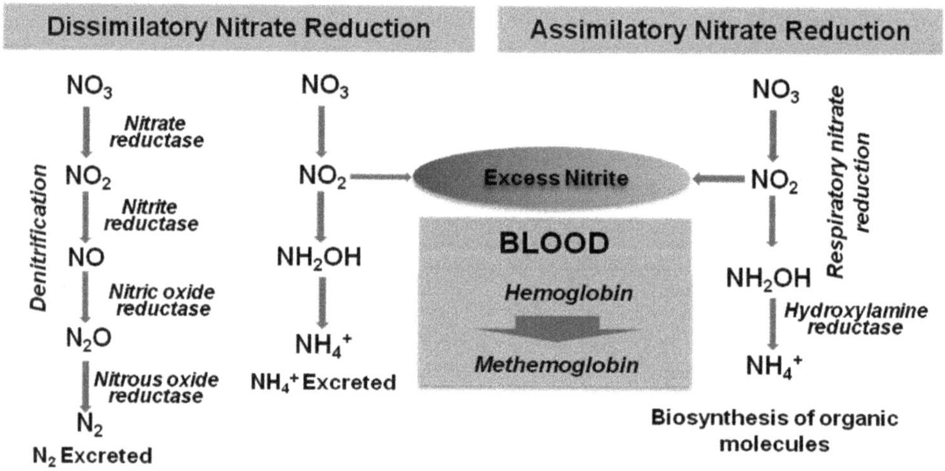

Figure 4. Assimilatory and dissimilatory routes of nitrate reduction.

Dissimilatory nitrate reduction to ammonia is the predominant pathway of nitrate metabolism in the rumen [80,81]. The conversion of nitrate to ammonia is thermodynamically more favorable to methanogenesis [67,81,150]. The reduction of nitrate to nitrite following reduction of nitrite to ammonia yields more Gibbs free energy than the reduction of CO_2 to CH_4 [67]. These processes could be the major route of H_2 elimination if sufficient nitrate is available in an actively fermenting rumen ecosystem. The conversion of nitrate to ammonia consumes eight reducing electrons and each mole of nitrate that is reduced could theoretically lower CH_4 production by one mole [80]. High organic matter concentration, low redox potential, and presence of sulfide in the rumen favors dissimilatory reduction [151], which is also not inhibited by high concentration of ammonia. The generated ammonia will be available for microbial biomass synthesis and provides an important supply of fermentable nitrogen [152]. The possibility of nitrate as an alternative H_2 sink to CO_2 in ruminant is somehow problematic and requires a detailed understanding of the rumen microbiome and the microbial processes involved in nitrate and nitrite metabolism due to the formation of toxic intermediates [147]. In the rumen, the nitrate conversion rate is higher than the rate of the subsequent conversion of nitrite to ammonia [139]. Excess nitrate consummation by the ruminant leads to the accumulation of nitrite which can trigger methemoglobinemia [82] and have adverse effects on the oxygen transport system due to oxidation of ferrous (Fe^{2+}) to the ferric (Fe^{3+}) yielding a stable oxidized form of hemoglobin (methemoglobin; MetHb) which is unable to release oxygen to the tissues. In mild cases, increases levels of MetHb can lower animal performance, but in severe cases this can be lethal [153].

A controlled and supervised administration of nitrate and nitrate-reducing bacteria into the rumen has been used as a successful strategy to allow the rumen and its microbiome to acclimatize to increasing levels of nitrate and enhance their ability to reduce nitrite [154,155]. Introducing microorganisms that have nitrite reductase activity and therefore an advantage over methanogenic archaea when competing for H_2 [156] ultimately affects ruminal CH_4 production and nitrite reduction [78,157]. *Veillonella parvula*, *Selenomonas ruminantium* and *Wolinella succinogenes* reduce nitrate and nitrite for example have been shown to be promising probiotcis that can be added to the rumen ecosystem to alleviate high nitrate concentrations in ruminant feed [83,158]. In another study, the occurrence of nitrate in diet controls nitrate-reducing bacteria *Wolinella succinogenes* and *Veillonella parvula* in medium containing ground hay and concentrate were estimated by competitive PCR [84]. Simon [79] studied the administration of *Wolinella succinogenes* having the ability to convert nitrate to ammonia with minimum nitrite accumulation in in vitro studies. A similar effect was also established in vivo using *Escherichia coli* strains with high nitrate/nitrate reductase activity [157]. The addition of anaerobic cultures of *E. coli* W3110 or *E. coli* nir-Ptac with nitrate in cultures of mixed ruminal microorganisms decrease nitrite toxicity and CH_4 production in vitro [159,160]. Sakthivel et al. [78] found that decreased CH_4 formation and enhanced nitrate and nitrite removal from ruminal digesta in the presence of a nitrate-reducing rumen bacterium (unidentified) in vitro. Nitrate supplementation linearly increased total VFA concentration and cellulolytic bacteria species (*Ruminococcus flavefaciens*, *Ruminococcus albus* and *Fibrobacter succinogenes*) in rumen-fistulated steers [85]. In the same study, they reported that *Campylobacter fetus*, *Selenomonas ruminantium*, and *Mannheimia succiniciproducens* were major nitrate-reducing bacteria in steers and their number linearly increased with level of nitrate supplementation [85]. Prebiotics also affect nitrate reduction and supplementation of galacto-oligosaccharide (GOS) decreased nitrite accumulation and up to 11% reduction in CH_4 emission [134,161].

Rumen protozoa have been reported to accelerate nitrate reduction when co-cultured with bacteria [162]. A significant proportion of nitrate reduction in the rumen with higher acetate to propionate ratio was observed with protozoa fraction in vitro [163]. van Zijderveld et al. [70] reported that though the number of protozoa remained unaffected with dietary supplementation of nitrate, yet a decline in protozoal count (60%) was observed with nitrate administration in the rumen [157]. Asanuma et al. [86] reported sevenfold declined protozoal population in goats adapted to dietary nitrate and significant decrease in the population of methanogen, protozoa, and fungi with increase in *S. bovis* and *S. ruminantium* was observed with nitrate supplementation. As methanogenic archaea are associated with protozoa surfaces, decrease in their number may have a crucial role in reducing CH_4 emission. Overall, inadequate information regarding the shift of rumen microbiome, in response to diet limits the challenges to reduce nitrate/nitrite reduction in rumen. Ruminant diet and substrate type also influence the reduction rate of nitrate and nitrite. In rumen bacterium *S. ruminantium* increase nitrate reductase per cell mass was reported in the presence of nitrate [164]. Higher reduction rate was observed on lactate as compared to glucose, and further enhanced with succinate. Regularly dosing nitrate directly into the rumen lowered CH_4 production was observed in vivo [87,165].

Leng [80] reported that a host of factors *viz.* fermentable carbohydrate, adequate sulfur level, low soluble protein fraction and a source of bypass protein favor the use of nitrate and lower CH_4 production. Hulshof et al. [88] found 32% decrease in CH_4 production in steers when fed nitrate at 2.2% of DM. van Zijderveld et al. [70] reported feeding nitrate or sulfate had no effect on the concentration of short chain fatty acids in rumen fluid after 24 h of feeding. However, the molar proportion of branched-chain VFAs varied, higher when sulfate and lower when nitrate fed diet were administered in lambs at a proportion of 2.6% of dry matter each in corn silage-based diet for 28 days. A significant decrease in CH_4 production was at maximum immediately after nitrate feeding, but the effect was uniform for the entire day in sulfate feeding. An increase in nitrate level in diet was accompanied by a linear decrease in CH_4 reduction [89]. Leng [80] observed inclusion of 1% potassium nitrate in a diet decreased CH_4 production by 10%. Decrease in CH_4 production by 23% in sheep was achieved with oat hay diet supplemented with 4% potassium nitrate compared to control diet made iso-nitrogenous by the addition of urea [90]. Nitrate supplementation has also been proposed to be a useful non-protein nitrogen (NPN) source for ruminants and as a replacement for urea [88,89,132,133,141,166]. Numerous in vivo and in vitro studies confirmed the efficacy of feeding nitrate on decreasing enteric CH_4 emissions without resulting in clinical signs of toxicity [70,89,90,150,166]. In an experiment when rumen fluid from a Jersey bull was incubated with sodium nitrate (12 mM) in vitro, 70% reduction of CH_4 level with 30% decrease in gas production was achieved [167]. Zhou et al. [91] further reported complete inhibition of CH_4 production with nitrate level more than 12 mM. The use of nitrate appears to be one of the improved strategies to be adapted in livestock sector to reduce enteric CH_4 fermentation, but the animals need to be acclimatized to nitrate feeding by step by step increasing the level in the diet to avoid harmful effects.

3.4. Propionogenesis

Redirection of metabolic H_2 away from CH_4 toward volatile fatty acids, primarily propionate, has been suggested to be an efficient strategy to reduce enteric CH_4 production in vivo, while also increasing the animal's feed efficiency [92,102,103,138,168,169]. The limiting factor for propionogenesis (Figure 5) as H_2 sink and a mean of consistently lowering the partial H_2 pressure in the rumen is substrate availability [63,170]. Formation of propionate can occur through either the microbial fumarate-succinate pathway [171] or the microbial conversion of pyruvate to lactate and acrylyl-CoA ester and the subsequent reduction in propionate [172]. Ellis et al. [50] reported that reduction of fumarate to succinate is thermodynamically more favorable than methanogenesis within the physiological partial H_2 pressure of the rumen that needs required substrate are availability in rumen.

In accordance with the notion that propionogenesis is a substrate-limited process, inclusion of propionate precursors or dicarboxylic acids in the diet shifted rumen fermentation toward propionic acid production and decreased CH_4 yield [93,94,173] (Table 2). The addition of fumaric acid for example yielded reduced in vitro and in vivo CH_4 production [53,91,94–98,140,174–177] and Wallace et al. [178] also observed an increase in weight gain in lambs fed when fumaric acid was added to the fed. Increased DM digestibility with decreased CH_4 production with addition of sodium fumarate in vitro was also observed [95]. Itabashi et al. [179] marked fumaric acid fed together with salinomycin to Holstein steers increased molar concentration of propionic acid resulting in a 16% decrease in CH_4 production, suggesting that the addition of ionophores together with fumarate might have a synergistic effect of these compounds on CH_4 production. In another study, steers fed with fumaric acid (2% DM) on a sorghum silage-based diet was reported to reduce CH_4 by 23% [96]. Asanuma et al. [140] suggested fumarate to be an economical feed additive for reduced CH_4 production. García-Martínez et al. [175] reported that effects of fumarate on rumen fermentation depend on the nature of the incubated substrate and significant response was observed with high forage diet. Wood et al. [135] reported CH_4 emission in sheep reduced to approximately 76% with fumaric acid (10%) encapsulated in fat. Demeyer and Henderickx [136] found 60% inhibition of CH_4 production by addition of fumarate (500 μM) in vitro. Similarly, a 17% decrease in CH_4 production in response to the addition of 400 μM fumarate was observed [94]. In another study, 38% reduction in CH_4 production in continuous fermenters was recorded with fumarate supplementation [137]. Ungerfeld et al. [176] recommended that low concentration of fumaric acid would be more effective in reducing CH_4 production. Beauchemin and McGinn [180] observed fumaric acid caused potential valuable changes in ruminal fermentation but no measurable reductions in CH_4 emissions. McGinn et al. [26] also reported that fumaric acid had no effect on CH_4 emissions, in growing beef cattle. Although the majority of the evidence support fumaric acid addition in animal diet, yet economic arguments and acidosis problems restrict their application in animal feed.

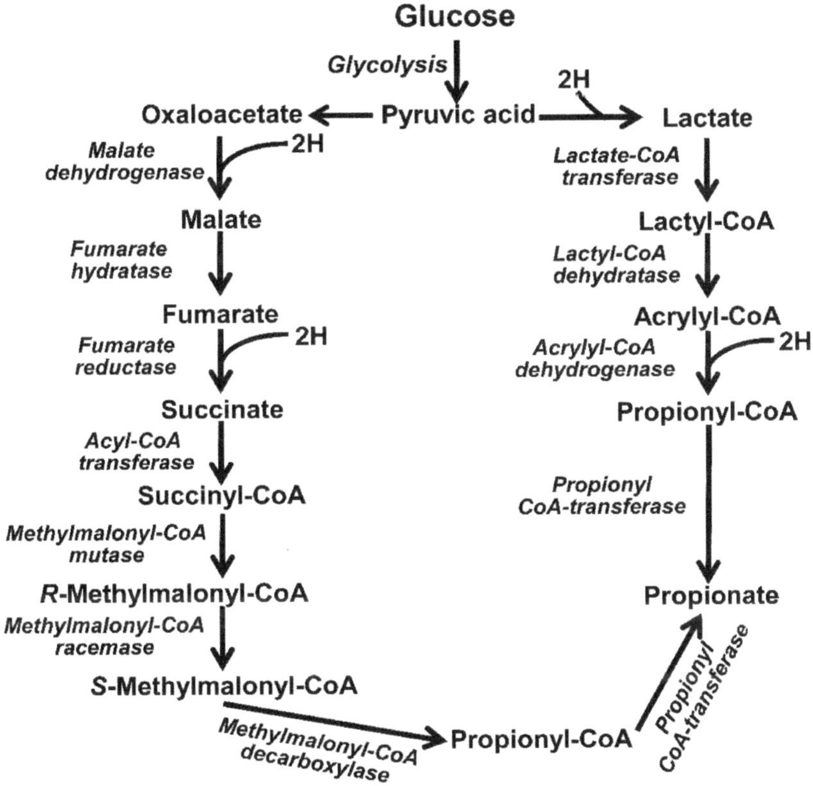

Figure 5. Pathway for propionate synthesis from oxaloacetate and lactate.

Similarly, the addition of malate, a key intermediate of the inverse citric acid cycle and of the succinate propionate pathway, has also been intensively studied for its ability to stimulate propionate production in the rumen [173,181,182]. Feed supplemented with malate (140 g/day) in lactating dairy cows reported an increase in milk production and feed conversion efficiency [183]. Dosing malate in ruminant diet increased nitrogen retention in sheep and steers while it also improved average daily weight gain and feed efficiency in calves [184]. In experiments, changes in rumen pH, VFA profiles and decreased CH_4 production analogous was also noticed when malate was added to diet [181,185,186]. Carro et al. [186] reported that although malate decreased CH_4 production per unit of DM digestion, but enhanced fiber digestibility resulted net increase in CH_4 production. Foley et al. [99,100] noticed little benefit gained from the dietary supplementation of malate in dairy cows. However, Carro and Ranilla [187,188] observed that malate beneficially affected in vitro rumen fermentation, with decreased CH_4 production and L-lactate concentrations.

In addition to the direct inclusion of propionate intermediates, several studies investigated the use of probiotics to enhance propionate production [57,93,101,140]. *Veillonella parvula, S. ruminantium* subsp. *ruminantium, S. ruminantium* subsp. *lactilytica* and *Fi-*

brobacter succinogenes have been reported to support propionate production and a high capacity to CH_4 reduction [189]. In another study, in vitro addition of fumarate-reducing bacteria *Mitsuokella jalaludinii* increased succinate production with significant decrease in CH_4 production and change in rumen microbial diversity was reported [190]. Nisbet and Martin [174] observed 10 mM-fumarate stimulated the growth of *S. ruminantium* in pure cultures. López et al. [95] found a significant increase in cellulolytic bacteria with addition of 7.35 mM-fumarate to semi-continuous fermenters. Similarly, Zhou et al. [101] observed addition of disodium fumarate inhibited the growth of methanogens, protozoa and fungi while cellulolytic bacteria (*R. albus*, *F. succinogenes* and *B. fibrisolvens*) and proteolytic bacteria (*B. fibrisolvens*, *P. ruminicola*, and *Clostridium* sp.) showed positive response. Compositional changes in the bacterial population in goat's rumen and improved metabolism of rumen lactate fermentation were also reported with addition of disodium fumarate [97]. In majority addition of fumarate and malate as well as microbe capable of reducing malate or fumarate seem to be effective in controlling ruminal methanogenesis and escalating supported microflora but dosing level and cost needs to be considered for this approach.

Addition of lactic acid bacteria (LAB) and their metabolites to enhance propionate synthesis for the H_2 sequestration also been invested by many researchers [141,191–193]. LAB stimulated the growth of lactate using microbes resulted in increased propionic acid production and leading to a substantial decrease in the H_2 availability for CH_4 production [141]. *Lactobacillus pentosus* D31 was reported to reduce CH_4 production (13%) over a period of 4 weeks dosed with 6×10^{10} cfu to each animal every day [194]. Bacteriocins, the metabolites of LAB reported to decrease CH_4 production with promising results both in vitro and in vivo experiments. The possibility of employing bacteriocins for CH_4 mitigation from streptococci of rumen origin has recently been reviewed [66]. Nisin a bacteriocin produced from *Lactococcus lactis* was observed to decrease 36% CH_4 production in vitro [195]. In sheep a 10% decrease in CH_4 emissions (g/kg DMI) with nisin supplementation was reported [196]. Similarly, bovicin HC5 [34] and pediocin [197] was shown to decrease CH_4 production by 53% and 49% in in vitro trials, respectively. LAB used as silage inoculants [198] also reported CH_4 diminishing (8.6%) activities and possible increase in propionic acid (4.8%). Reduced gas production by silage treated groups compared with the untreated silage evoked a shift in fermentation [199]. Similarly, in another experiment carried out by Cao et al. [200] with vegetable residue silage, there was a decrease in CH_4 production (46.6% reduction) and increase in in vitro dry matter digestibility. Huyen et al. [201] reported that LAB strains are most promising when used as silage inoculants and observed to decrease CH_4 production and increase DM digestibility. In addition to that increase in cellulolytic microorganisms and decrease in CH production using corn stover silage inoculated with *Lactobacillus plantarum* and increase in lactic acid fermentation, in vitro digestibility and CH_4 mitigation in the forage sorghum mixture silages was also observed using *Lactobacillus casei* TH14 inoculant [202]. These experiments showed that LAB, their metabolites, and applications in silage have a positive effect in decreasing CH_4 yield possibly by stimulating the lactate-propionate pathway or release of inhibitory compounds.

4. Conclusions and Future Prospects

Ruminal methanogenesis, contributing CH_4 to the atmosphere, is directly and inversely linked to the animal productivity. The ability to control CH_4 emission especially reduce methanogenesis from agriculture has enormous environmental and socioeconomic implication, but it also requires a detailed understanding of the microorganisms and microbial processes that are involved. Although a complete understanding of these highly interwoven microbial and metabolic networks has still not been achieved and most likely will not be feasible in the immediate future, there are some aspects that are reasonably well understood. These aspects represent a promising starting point for targeted CH_4 reduction from ruminants. One of the promising key intermediates that has been recognized as such and that has received significant attention for targeted CH_4 mitigation is metabolic H_2 and the metabolic pathways, microbes and enzymes involved its production and consumption.

Since H_2 is an immediate precursor for the archaeal reduction of CO_2 into CH_4, biological approaches that redirect H_2 away from archaeal methanogenesis and into alternate metabolic pathways seem to be the most promising approaches to convert feed carbon into metabolic energy for the ruminant instead of releasing it into the atmosphere. Redirecting H_2 through reductive acetogenesis and propionogenesis has advantages over other pathways due to production of valuable metabolic end products that can be used by the host animal as nutrients and can be converted into animal proteins for human consumption. Although our understanding of how to redirect metabolic H_2 into more favorable pathways facilitates the production of value-added metabolic intermediates and therefore redirects otherwise lost feed energy, several issues related to the fine tuning of this redirection, such as the co-factor requirements, toxicity of metabolic intermediates, as well as thermodynamics of competing metabolic processes, need to be investigated in greater detail. A further aspect that will have to be investigated further and that will have direct implications for the translational value of findings on the area of rumen nutrition and function is the link and dependence of the rumen microbiome and its function in dietary conversion. With recent advances in omics technologies and the foray into the metabolic processes that are actually engaged in the rumen microbiome under certain physiological conditions, we now have the tools that will enable us to lay the foundation for a high-resolution picture of the rumen ecosystem and its microbial processes.

Author Contributions: P.K.C. conceived the idea for the article, prepared and edited the final manuscript. R.J. and P.K.C. prepared the figure, tables and edited the final manuscript. A.K.P. and S.K.T. help in final editing and participated in developing the idea and critically revised the manuscript. All authors have read and agreed to the published version of the manuscript.

Funding: This research received no external funding.

Institutional Review Board Statement: Not applicable.

Informed Consent Statement: Not applicable.

Data Availability Statement: Not applicable.

Acknowledgments: The authors would like to acknowledge for necessary facilities provided by ICAR-National Dairy Research Institute, Karnal, Haryana, India for literature collection of this review. Institutional fellowship received to PKC during the period of study is highly obliged.

Conflicts of Interest: The authors declare no conflict of interest.

References

1. Wang, S.; Giller, K.; Kreuzer, M.; Ulbrich, S.E.; Braun, U.; Schwarm, A. Contribution of ruminal fungi, archaea, protozoa, and bacteria to the methane suppression caused by oilseed supplemented diets. *Front. Microbiol.* **2017**, *8*, 1864. [CrossRef]
2. Choudhury, P.K.; Salem, A.Z.M.; Jena, R.; Kumar, S.; Singh, R.; Puniya, A.K. Rumen microbiology: An overview. In *Rumen Microbiology-Evolution to Revolution*; Puniya, A.K., Singh, R., Kamra, D.N., Eds.; CRC Springer: New Delhi, India, 2015; pp. 3–16.
3. Kumar, S.; Dagar, S.S.; Puniya, A.K.; Upadhyay, R.C. Changes in methane emission, rumen fermentation in response to diet and microbial interactions. *Res. Vet. Sci.* **2013**, *94*, 263–268. [CrossRef] [PubMed]
4. Kamra, D.N. Rumen microbial ecosystem. *Curr. Sci.* **2005**, *89*, 124–135.
5. Janssen, P.H.; Kirs, M. Structure of the archaeal community of the rumen. *Appl. Environ. Microbiol.* **2008**, *74*, 3619–3625. [CrossRef] [PubMed]
6. Kumar, S.; Choudhury, P.K.; Carro, M.D.; Griffith, G.W.; Dagar, S.S.; Puniya, M.; Calabro, S.; Ravella, S.R.; Dhewa, T.; Upadhyay, R.C.; et al. New aspects and strategies for methane mitigation from ruminants. *Appl. Microbiol. Biotechnol.* **2014**, *98*, 31–44. [CrossRef]
7. Ragsdale, S.W.; Pierce, E. Acetogenesis and the Wood-Ljungdahl pathway of CO(2) fixation. *Biochim. Biophys. Acta* **2008**, *1784*, 1873–1898. [CrossRef]
8. Ishaq, S.L.; AlZahal, O.; Walker, N.; McBride, B. An investigation into rumen fungal and protozoal diversity in three rumen fractions, during high-fiber or grain-induced sub-acute ruminal acidosis conditions, with or without active dry yeast supplementation. *Front. Microbiol.* **2017**, *8*, 1943. [CrossRef]

9. Malik, P.K.; Bhatta, R.; Gagen, E.J.; Sejian, V.; Soren, N.M.; Prasad, C.S. Alternate H_2 sinks for reducing rumen methanogenesis. In *Climate Change Impact on Livestock: Adaptation and Mitigation*; Sejian, V., Gaughan, J., Baumgard, L., Prasad, C., Eds.; Springer: New Delhi, India, 2015; pp. 303–320.
10. Joblin, K. Methanogenic archaea. In *Methods in Gut Microbial Ecology for Ruminants*; Makker, H., McSweeney, C.S., Eds.; Springer: Dordrecht, The Netherlands, 2005; pp. 47–53.
11. Moss, A.R.; Jouany, J.P.; Newbold, J. Methane production by ruminants: Its contribution to global warming. *Ann. Zootech.* **2000**, *49*, 231–253. [CrossRef]
12. Ungerfeld, E.M. Inhibition of rumen methanogenesis and ruminant productivity: A meta-analysis. *Front. Vet. Sci.* **2018**, *5*, 113. [CrossRef]
13. Beauchemin, K.A.; Ungerfeld, E.M.; Eckard, R.J.; Wang, M. Review: Fifty years of research on rumen methanogenesis: Lessons learned and future challenges for mitigation. *Animal* **2020**, *14*, s2–s16. [CrossRef]
14. Van Nevel, C.J.; Demeyer, D.I. Control of rumen methanogenesis. *Environ. Monit. Assess.* **1996**, *42*, 73–97. [CrossRef] [PubMed]
15. Demeyer, D.; Fiedler, D.; De Graeve, K.G. Attempted induction of acetogenesis into the rumen fermentation in vitro. *Reprod. Nutr. Dev.* **1996**, *36*, 233–240. [CrossRef] [PubMed]
16. Garsa, A.K.; Choudhury, P.K.; Puniya, A.K.; Dhewa, T.; Malik, R.K.; Tomar, S.K. Bovicins: The bacteriocins of streptococci and their potential in methane mitigation. *Probiotics Antimicrob. Proteins* **2019**, *11*, 1403–1413. [CrossRef]
17. Goopy, J.P. Creating a low enteric methane emission ruminant: What is the evidence of success to the present and prospects for developing economies? *Anim. Prod. Sci.* **2019**, *59*, 1769–1776. [CrossRef]
18. Hook, S.E.; Wright, A.D.G.; McBride, B.W. Methanogens: Methane producers of the rumen and mitigation strategies. *Archaea* **2010**, *2010*, 945785. [CrossRef]
19. Kim, S.H.; Mamuad, L.L.; Islam, M.; Lee, S.S. Reductive acetogens isolated from ruminants and their effect on in vitro methane mitigation and milk performance in Holstein cows. *J. Anim. Sci. Technol.* **2020**, *62*, 1–13. [CrossRef]
20. Martin, C.; Morgavi, D.P.; Doreau, M. Methane mitigation in ruminants: From microbe to the farm scale. *Animal* **2010**, *4*, 351–365. [CrossRef]
21. Wanapat, M.; Kongmun, P.; Poungchompu, O.; Cherdthong, A.; Khejornsart, P.; Pilajun, R.; Kaenpakdee, S. Effects of plants containing secondary compounds and plant oils on rumen fermentation and ecology. *Trop. Anim. Health Prod.* **2012**, *44*, 399–405. [CrossRef]
22. Beauchemin, K.A.; Kreuzer, M.O.; O'Mara, F.P.; McAllister, T.A. Nutritional management for enteric methane abatement: A review. *Aust. J. Exp. Agric.* **2008**, *48*, 21–27. [CrossRef]
23. García-González, R.; López, S.; Fernández, M.; Bodas, R.; González, J.S. Screening the activity of plants and spices for decreasing ruminal methane production in vitro. *Anim. Feed Sci. Technol.* **2008**, *147*, 36–52. [CrossRef]
24. Eckard, R.J.; Grainger, C.; de Klein, C.A.M. Options for the abatement of methane and nitrous oxide from ruminant production: A review. *Livest. Sci.* **2011**, *130*, 47–56. [CrossRef]
25. Patra, A.; Park, T.; Kim, M.; Yu, Z. Rumen methanogens and mitigation of methane emission by anti-methanogenic compounds and substances. *J. Anim. Sci. Biotechnol.* **2017**, *8*, 13. [CrossRef] [PubMed]
26. McGinn, S.M.; Beauchemin, K.A.; Coates, T.; Colombatto, D. Methane emissions from beef cattle: Effects of monensin, sunflower oil, enzymes, yeast and furmaric acid. *J. Anim. Sci.* **2004**, *82*, 3346–3356. [CrossRef] [PubMed]
27. Janssen, P.H.; Frenzel, P. Inhibition of methanogenesis by methyl fluoride: Studies of pure and defined mixed cultures of anaerobic bacteria and archaea. *Appl. Environ. Microbiol.* **1997**, *63*, 4552–4557. [CrossRef]
28. Miller, T.L.; Wolin, M.J. Inhibition of growth of methane-producing bacteria of the ruminant forestomach by hydroxymethylglutaryl-SCoA reductase inhibitors. *J. Dairy Sci.* **2001**, *84*, 1445–1448. [CrossRef]
29. Mitsumori, M.; Shinkai, T.; Takenaka, A.; Enishi, O.; Higuchi, K.; Kobayashi, Y.; Nonaka, I.; Asanuma, N.; Denman, S.E.; McSweeney, C.S. Responses in digestion, rumen fermentation and microbial populations to inhibition of methane formation by a halogenated methane analogue. *Br. J. Nutr.* **2012**, *108*, 482–491. [CrossRef]
30. Sanchez, J.M.; Valle, L.; Rodriguez, F.; Morinnigo, M.A.; Borrego, J.J. Inhibition of methanogenesis by several heavy metals using pure cultures. *Lett. Appl. Microbiol.* **1996**, *23*, 439–444. [CrossRef]
31. Fievez, V.F.; Dohme, M.; Daneels, K.R.; Demeyer, D. Fish oils as potent rumen methane inhibitors and associated effects on rumen fermentation in vitro and in vivo. *Anim. Feed Sci. Technol.* **2003**, *104*, 41–58. [CrossRef]
32. Machmüller, A.; Kreuzer, M. Methane suppression by coconut oil and associated effects on nutrient and energy balance in sheep. *Can. J. Anim. Sci.* **1999**, *79*, 65–72. [CrossRef]
33. Machmüller, A. Medium-chain fatty acids and their potential to reduce methanogenesis in domestic ruminants. *Agric. Ecosyst. Environ.* **2006**, *112*, 107–114. [CrossRef]
34. Lee, S.S.; Hsu, J.T.; Mantovani, H.C.; Russell, J.B. The effect of bovicin HC5, a bacteriocin from *Streptococcus bovis* HC5, on ruminal methane production in vitro. *FEMS Microbiol. Lett.* **2002**, *217*, 51–55. [CrossRef]
35. Baker, S.K. Rumen methanogens and inhibition of methanogenesis. *Aust. J. Agric. Res.* **1999**, *50*, 1293–1298. [CrossRef]
36. Anderson, R.C.; Carstens, G.E.; Miller, R.K.; Callaway, T.R.; Schultz, C.L.; Edrington, T.S.; Harvey, R.; Nisbet, D. Effect of oral nitroethane and 2-nitropropanol administration on methane-producing activity and volatile fatty acid production in the ovine rumen. *Bioresour. Technol.* **2006**, *97*, 2421–2426. [CrossRef] [PubMed]

37. Wright, A.D.G.; Kennedy, P.; O'neill, C.J.; Toovey, A.F.; Popovski, S.; Rea, S.M.; Pimm, C.L.; Klein, L. Reducing methane emissions in sheep by immunization against rumen methanogens. *Vaccine* **2004**, *22*, 3976–3985. [CrossRef] [PubMed]
38. Zhang, L.; Huang, X.; Xue, B.; Peng, Q.; Wang, Z.; Yan, T.; Wang, L. Immunization against rumen methanogenesis by vaccination with a new recombinant protein. *PLoS ONE* **2015**, *10*, e0140086. [CrossRef]
39. Newbold, C.J.; de la Fuente, G.; Belanche, A.; Ramos-Morales, E.; McEwan, N.R. The Role of Ciliate Protozoa in the Rumen. *Front. Microbiol.* **2015**, *6*, 1313. [CrossRef]
40. LeVan, T.D.; Robinson, J.A.; John, R.; Greening, R.C.; Smolenski, W.J.; Leedle, J.A.Z.; Schaefer, D.M. Assessment of reductive acetogenesis with indigenous ruminal bacterium populations and *Acetitomaculum ruminis*. *Appl. Environ. Microbiol.* **1998**, *64*, 3429–3436. [CrossRef]
41. Nollet, L.; Demeyer, D.; Verstraete, W. Effect of 2-bromoethanesulfonic acid and *Peptostreptococcus productus* ATCC35244 addition on stimulation of reductive acetogenesis in the ruminal ecosystem by selective inhibition of methanogenesis. *Appl. Environ. Microbiol.* **1997**, *63*, 194–200. [CrossRef]
42. Nollet, L.; Velde, I.V.; Verstraete, W. Effect of the addition of *Peptostreptococcus productus* ATCC35244 on the gastro intestinal microbiota and its activity, as simulated in an in vitro simulator of the human gastro-intestinal tract. *Appl. Microbiol. Biotechnol.* **1997**, *48*, 99–104. [CrossRef]
43. Liu, H.; Wang, J.; Wang, A.; Chen, J. Chemical inhibitors of methanogenesis and putative applications. *Appl. Microbiol. Biotechnol.* **2011**, *89*, 1333–1340. [CrossRef]
44. Hungate, R.E. *The Rumen and Its Microbes*; Academic Press: New York, NY, USA, 1966; p. 533.
45. Ungerfeld, E.M. Metabolic hydrogen flows in rumen fermentation: Principles and possibilities of interventions. *Front. Microbiol.* **2020**, *11*, 589. [CrossRef] [PubMed]
46. Williams, A.G.; Coleman, G.S. The rumen protozoa. In *The Rumen Microbial Ecosystem*, 2nd ed.; Hobson, P.N., Stewarteds, C.S., Eds.; Blackie Academic & Professional: New York, NY, USA, 1997; pp. 73–139.
47. Cord-Ruwisch, R.; Seitz, H.J.; Conrad, R. The capacity of hydrogenotrophic anaerobic bacteria to compete for traces of hydrogen depends on the redox potential of the electron acceptor. *Arch. Microbiol.* **1988**, *149*, 350–357. [CrossRef]
48. Immig, I. The rumen and hindgut as a source of ruminant methanogenesis. *Environ. Monit. Assess.* **1996**, *42*, 57–72. [CrossRef] [PubMed]
49. McAllister, T.A.; Newbold, C.J. Redirecting rumen fermentation to reduce methanogenesis. *Aust. J. Exp. Agric.* **2008**, *48*, 7–13. [CrossRef]
50. Ellis, J.L.; Dijkstra, J.; Kebreab, E.; Bannink, A.; Odongo, N.E.; Mcbride, B.W.; France, J. Aspects of rumen microbiology central to mechanistic modelling of methane production in cattle. *J. Agric. Sci.* **2008**, *146*, 213–233. [CrossRef]
51. Hegarty, R.S.; Gerdes, R. Hydrogen production and transfer in the rumen. *Rec. Adv. Anim. Nutr.* **1999**, *12*, 37–44. Available online: http://livestocklibrary.com.au/handle/1234/19891 (accessed on 25 October 2022).
52. Greening, C.; Geier, R.; Wang, C.; Woods, L.C.; Morales, S.E.; McDonald, M.J.; Rushton-Green, R.; Morgan, X.C.; Koike, S.; Leahy, S.C.; et al. Diverse hydrogen production and consumption pathways influence methane production in ruminants. *ISME J.* **2019**, *13*, 2617–2632. [CrossRef]
53. Morgavi, D.P.; Forano, E.; Martin, C.; Newbold, C.J. Microbial ecosystem and methanogenesis in ruminants. *Animal* **2010**, *4*, 1024–1036. [CrossRef]
54. Gagen, E.J.; Denman, S.E.; McSweeney, C.S. Acetogenesis as an alternative to methanogenesis in the rumen. In *Livestock Production and Climate Change*; Malik, P.K., Bhatta, R., Takahashi, J., Kohn, R.A., Prasad, C.S., Eds.; CABI: Wallingford, UK, 2015; pp. 292–303.
55. Morvan, B.; Bonnemoy, F.; Fonty, G.; Gouet, P. Quantitative determination of H_2-utilizing acetogenic and sulfate-reducing bacteria and methanogenic archaea from digestive tract of different mammals. *Curr. Microbiol.* **1996**, *32*, 129–133. [CrossRef]
56. Kim, S.H.; Mamuad, L.L.; Kim, D.W.; Kim, S.K.; Lee, S.S. Fumarate reductase-producing enterococci reduce methane production in rumen fermentation in vitro. *J. Microbiol. Biotechnol.* **2016**, *26*, 558–566. [CrossRef]
57. Mamuad, L.L.; Kim, S.H.; Lee, S.S.; Cho, K.K.; Jeon, C.O.; Lee, S.S. Characterization, metabolites and gas formation of fumarate reducing bacteria isolated from Korean native goat (*Capra hircus* coreanae). *J. Microbiol.* **2012**, *50*, 925–931. [CrossRef] [PubMed]
58. Wolin, M.J.; Miller, T.L.; Stewart, C.S. Microbe-microbe interactions. In *The Rumen Microbial Ecosystem*; Hobson, P.N., Stewart, C.S., Eds.; Chapman and Hall: London, UK, 1997; pp. 467–488.
59. Joblin, K.N. Ruminal acetogens and their potential to lower ruminant methane emissions. *Aust. J. Agric. Res.* **1999**, *50*, 1307–1313. [CrossRef]
60. Ljungdahl, L.G. The autotrophic pathway of acetate synthesis in acetogenic bacteria. *Annu. Rev. Microbiol.* **1986**, *40*, 415–450. [CrossRef] [PubMed]
61. Morvan, B.; Doré, J.; Rieu-Lesme, F.; Foucat, L.; Fonty, G.; Gouet, P. Establishment of hydrogen-utilizing bacteria in the rumen of the newborn lamb. *FEMS Microbiol. Lett.* **1994**, *117*, 249–256. [CrossRef] [PubMed]
62. Ragsdale, S.W. Enzymology of the Wood Ljungdahl pathway of acetogenesis. *Ann. N. Y. Acad. Sci.* **2008**, *1125*, 129–136. [CrossRef]
63. Attwood, G.; McSweeney, C. Methanogen genomics to discover targets for methane mitigation technologies and options for alternative hydrogen utilization in the rumen. *Aust. J. Exp. Agric.* **2008**, *48*, 28–37. [CrossRef]

64. Drake, H.L.; Küsel, K.; Matthies, C. Acetogenic prokaryotes. In *The Prokaryotes: A Handbook on the Biology of Bacteria: Ecophysiology and Biochemistry*; Dworkin, M., Stanley, F., Rosenberg, E., Schleifer, K.H., Stackebrandt, E., Eds.; Springer: New York, NY, USA, 2006; pp. 354–420.
65. Drake, H.L.; Gössner, S.A.; Daniel, S.L. Old acetogens, new light. *Ann. N. Y. Acad. Sci.* **2008**, *1125*, 100–128. [CrossRef]
66. Klieve, A.V.; Ouwerkerk, D. Comparative greenhouse gas emissions from herbivores. In *Proceedings of the VII International Symposium on the Nutrition of Herbivores*; Meng, Q.X., Ed.; China Agricultural University Press: Beijing, China, 2007; pp. 17–21.
67. Ungerfeld, E.M.; Kohn, R.A. The role of thermodynamics in the control of ruminal fermentation. In *Ruminant Physiology: Digestion, Metabolism and Impact of Nutrition on Gene Expression, Immunology and Stress*; Sejrsen, K., Hvelplund, T., Nielsen, M.O., Eds.; Wageningen Academic Publishers: Wageningen, The Netherlands, 2006; pp. 55–85.
68. Breznak, J.A.; Switzer, J.M. Acetate synthesis from H_2 plus CO_2 by termite gut microbes. *Appl. Environ. Microbiol.* **1986**, *52*, 623–630. [CrossRef]
69. Breznak, J.A.; Kane, M.D. Microbial H_2/CO_2 acetogenesis in animal guts: Nature and nutritional significance. *FEMS Microbiol. Rev.* **1990**, *87*, 309–314. [CrossRef]
70. Van Zijderveld, S.M.; Gerrits, W.J.; Apajalahti, J.A.; Newbold, J.R.; Dijkstra, J.; Leng, R.A.; Perdok, H. Nitrate and sulfate: Effective alternative hydrogen sinks for mitigation of ruminal methane production in sheep. *J. Dairy Sci.* **2010**, *93*, 5856–5866. [CrossRef]
71. Silivong, P.; Preston, T.R.; Leng, R.A. Effect of sulphur and calcium nitrate on methane production by goats fed a basal diet of molasses supplemented with mimosa (*Mimosa pigra*) foliage. *Livest. Res. Rural Dev.* **2011**, *23*, 3. Available online: http://www.lrrd.org/lrrd23/3/sili23058.htm (accessed on 25 October 2022).
72. Wu, H.; Meng, Q.; Yu, Z. Effect of pH buffering capacity and sources of dietary sulfur on rumen fermentation, sulfide production, methane production, sulfate reducing bacteria, and total Archaea in in vitro rumen cultures. *Bioresour. Technol.* **2015**, *186*, 25–33. [CrossRef] [PubMed]
73. Binversie, E.Y.; Ruiz-Moreno, M.; Carpenter, A.; Heins, B.; Crawford, G.; DiCostanzo, A.; Stern, M. Effects of dietary roughage and sulfur in diets containing corn dried distillers' grains with solubles on hydrogen sulfide production and fermentation by rumen microbes in vitro. *J. Anim. Sci.* **2016**, *94*, 3883–3893. [CrossRef] [PubMed]
74. Shah, A.M.; Ma, J.; Wang, Z.; Hu, R.; Wang, X.; Peng, Q.; Amevor, F.K.; Goswami, N. Production of hydrogen sulfide by fermentation in rumen and its impact on health and production of animals. *Processes* **2020**, *8*, 1169. [CrossRef]
75. Drewnoski, M.E.; Richter, E.L.; Hansen, S.L. Dietary sulfur concentration affects rumen hydrogen sulfide concentrations in feedlot steers during transition and finishing. *J. Anim. Sci.* **2012**, *90*, 4478–4486. [CrossRef]
76. Gagen, E.J.; Mosoni, P.; Denman, S.E.; Al Jassim, R.; McSweeney, C.S.; Forano, E. Methanogen colonisation does not significantly alter acetogen diversity in lambs isolated 17 h after birth and raised aseptically. *Microb. Ecol.* **2012**, *64*, 628–640. [CrossRef]
77. Fonty, G.; Joblin, K.; Chavarot, M.; Roux, R.; Naylor, G.; Michallon, F. Establishment and development of ruminal hydrogenotrophs in methanogen-free lambs. *Appl. Environ. Microbiol.* **2007**, *73*, 6391–6403. [CrossRef]
78. Sakthivel, P.C.; Kamra, D.N.; Agarwal, N.; Chaudhary, L.C. Effect of sodium nitrate and nitrate reducing bacteria on in vitro methane production and fermentation with buffalo rumen liquor. *Asian-Australas. J. Anim. Sci.* **2012**, *25*, 812–817. [CrossRef]
79. Simon, J. Enzymology and bioenergetics of respiratory nitrite ammonification. *FEMS Microbiol. Rev.* **2002**, *26*, 285–309. [CrossRef]
80. Leng, R.A. The potential of feeding nitrate to reduce enteric methane production in ruminants. In *Report: The Department of Climate Change*; Commonwealth Government of Australia: Canberra, Australia, 2008.
81. Takahashi, J. Some prophylactic options to mitigate methane emission from animal agriculture in Japan. *Asian-Australas. J. Anim. Sci.* **2011**, *24*, 285–294. [CrossRef]
82. Van Zijderveld, S.M.; Fonken, B.; Dijkstra, J.; Gerrits, W.J.; Perdok, H.B.; Fokkink, W.; Newbold, J. Effects of a combination of feed additives on methane production, diet digestibility, and animal performance in lactating dairy cows. *J. Dairy Sci.* **2011**, *94*, 1445–1454. [CrossRef] [PubMed]
83. Yang, C.; Rooke, J.A.; Cabeza, I.; Wallace, R.J. Nitrate and inhibition of ruminal methanogenesis: Microbial ecology, obstacles, and opportunities for lowering methane emissions from ruminant livestock. *Front. Microbiol.* **2016**, *7*, 132. [CrossRef]
84. Iwamoto, M.; Asanuma, N.; Hino, T. Ability of *Selenomonas ruminantium*, *Veillonella parvula* and *Wolinella succinogenes* to reduce nitrate and nitrite with special reference to the suppression of ruminal methanogenesis. *Anaerobe* **2002**, *8*, 209–215. [CrossRef]
85. Zhao, L.; Meng, Q.; Ren, L.; Liu, W.; Zhang, X.; Huo, Y.; Zhou, Z. Effects of nitrate addition on rumen fermentation, bacterial biodiversity and abundance. *Asian-Australas. J. Anim. Sci.* **2015**, *28*, 1433–1441. [CrossRef] [PubMed]
86. Asanuma, N.; Yokoyama, S.; Hino, T. Effects of nitrate addition to a diet on fermentation and microbial populations in the rumen of goats, with special reference to *Selenomonas ruminantium* having the ability to reduce nitrate and nitrite. *Anim. Sci. J.* **2015**, *86*, 378–384. [CrossRef]
87. Sar, C.; Santoso, B.; Mwenya, B.; Gamo, Y.; Kobayashi, T.; Morikawa, R.; Kimura, K.; Mizukoshi, H.; Takahashi, J. Manipulation of rumen methanogenesis by the combination of nitrate with β 1-4 galacto-oligosaccharides or nisin in sheep. *Anim. Feed Sci. Technol.* **2004**, *115*, 129–142. [CrossRef]
88. Hulshof, R.B.; Berndt, A.; Gerrits, W.J.; Dijkstra, J.; van Zijderveld, S.M.; Newbold, J.R.; Perdok, H.B. Dietary nitrate supplementation reduces methane emission in beef cattle fed sugarcane-based diets. *J. Anim. Sci.* **2012**, *90*, 2317–2323. [CrossRef]
89. Van Zijderveld, S.M.; Gerrits, W.J.; Dijkstra, J.; Newbold, J.R.; Hulshof, R.B.; Perdok, H.B. Persistency of methane mitigation by dietary nitrate supplementation in dairy cows. *J. Dairy Sci.* **2011**, *94*, 4028–4038. [CrossRef]

90. Nolan, J.V.; Hegarty, R.S.; Hegarty, J.; Godwin, I.R.; Woodgate, R. Effects of dietary nitrate on fermentation, methane production and digesta kinetics in sheep. *Anim. Prod. Sci.* **2010**, *50*, 801–806. [CrossRef]
91. Zhou, Z.; Yu, Z.; Meng, Q. Effects of nitrate on methane production, fermentation, and microbial populations in in vitro ruminal cultures. *Bioresour. Technol.* **2012**, *103*, 173–179. [CrossRef]
92. Janssen, P.H. Influence of hydrogen on rumen methane formation and fermentation balances through microbial growth kinetics and fermentation thermodynamics. *Anim. Feed Sci. Technol.* **2010**, *160*, 1–22. [CrossRef]
93. Chen, J.; Harstad, O.M.; McAllister, T.; Dörsch, P.; Holo, H. Propionic acid bacteria enhance ruminal feed degradation and reduce methane production in vitro. *Acta Agric. Scand. Sect. A-Anim. Sci.* **2020**, *69*, 169–175. [CrossRef]
94. Newbold, C.J.; López, S.; Nelson, N.; Ouda, J.O.; Wallace, R.J.; Moss, A.R. Propionate precursors and other metabolic intermediates as possible alternative electron acceptors to methanogenesis in ruminal fermentation in vitro. *Br. J. Nutr.* **2005**, *94*, 27–35. [CrossRef] [PubMed]
95. López, S.; Valde's, C.; Newbold, C.J.; Wallace, R.J. Influence of sodium fumarate addition on rumen fermentation in vitro. *Br. J. Nutr.* **1999**, *81*, 59–64. [CrossRef]
96. Bayaru, E.; Kanda, S.; Kamada, T.; Itabashi, H.; Andoh, S.; Nishida, T.; Ishida, M.; Itoh, T.; Nagara, K.; Isobe, Y. Effect of fumaric acid on methane production, rumen fermentation and digestibility of cattle fed roughage alone. *Anim. Sci. J.* **2001**, *72*, 139–146. [CrossRef]
97. Mao, S.Y.; Zhang, G.; Zhu, W.Y. Effect of disodium fumarate on ruminal metabolism and rumen bacterial communities as revealed by denaturing gradient gel electrophoresis analysis of 16S ribosomal DNA. *Anim. Feed Sci. Technol.* **2008**, *140*, 293–306. [CrossRef]
98. Lin, B.; Lu, Y.; Salem, A.Z.M.; Wang, J.H.; Liang, Q.; Liu, J.X. Effects of essential oil combinations on sheep ruminal fermentation and digestibility of a diet with fumarate included. *Anim. Feed Sci. Technol.* **2013**, *184*, 24–32. [CrossRef]
99. Foley, P.A.; Kenny, D.A.; Lovett, D.K.; Callan, J.J.; Boland, T.M.; O'Mara, F.P. Effect of DL-malic acid supplementation on feed intake, methane emissions, and performance of lactating dairy cows at pasture. *J. Dairy Sci.* **2009**, *92*, 3258–3264. [CrossRef]
100. Foley, P.; Kenny, D.; Callan, J.; Boland, T.; O'mara, F. Effect of DL-malic acid supplementation on feed intake, methane emission, and rumen fermentation in beef cattle. *J. Anim. Sci.* **2009**, *87*, 1048–1057. [CrossRef] [PubMed]
101. Zhou, Y.W.; McSweeney, C.S.; Wang, J.K.; Liu, J.X. Effects of disodium fumarate on ruminal fermentation and microbial communities in sheep fed on high-forage diets. *Animal* **2012**, *6*, 815–823. [CrossRef]
102. Bannink, A.; Kogut, J.; Dijkstra, J.; France, J.; Kebreab, E.; Van Vuuren, A.M.; Tamminga, S. Estimation of the stoichiometry of volatile fatty acid production in the rumen of lactating cows. *J. Theor. Biol.* **2006**, *238*, 36–51. [CrossRef] [PubMed]
103. Ungerfeld, E.M. A theoretical comparison between two ruminal electron sinks. *Front. Microbiol.* **2013**, *4*, 319. [CrossRef] [PubMed]
104. Mackie, R.I.; Bryant, M.P. Acetogenesis and the rumen: Syntrophic relationships. In *Acetogenesis*; Drake, H.L., Ed.; Chapman and Hall: New York, NY, USA, 1994; pp. 331–364.
105. Pinder, R.S.; Patterson, J.A. Growth of acetogenic bacteria in response to varying pH, acetate or carbohydrate concentration. *Agric. Food Anal. Bacteriol.* **2012**, *3*, 6–16.
106. López, S.; McIntosh, F.M.; Wallace, R.J.; Newbold, C.J. Effect of adding acetogenic bacteria on methane production by mixed rumen microorganisms. *Anim. Feed Sci. Technol.* **1999**, *78*, 1–9. [CrossRef]
107. Mitsumori, M.; Matsui, H.; Tajima, K.; Shinkai, T.; Takenaka, A.; Denman, S.E.; McSweeney, C.S. Effect of bromochloromethane and fumarate on phylogenetic diversity of the formyltetrahydrofolate synthetase gene in bovine rumen. *Anim. Sci. J.* **2014**, *85*, 25–31. [CrossRef]
108. Faichney, G.J.; Graham, N.M.; Walker, D.M. Rumen characteristics, methane emission, and digestion in weaned lambs reared in isolation. *Aust. J. Agric. Res.* **1999**, *50*, 1083–1090. [CrossRef]
109. Chaucheyras, F.; Fonty, G.; Bertin, G.; Gouet, P. In vitro H_2 utilization by a ruminal acetogenic bacterium cultivated alone or in association with an archaea methanogen is stimulated by a probiotic strain of *Saccharomyces cerevisiae*. *Appl. Environ. Microbiol.* **1995**, *61*, 3466–3467. [CrossRef]
110. Yang, C.; Guan, L.; Liu, J.; Wang, J. Rumen fermentation and acetogen population changes in response to an exogenous acetogen TWA4 strain and *Saccharomyces cerevisiae* fermentation product. *J. Zhejiang Univ. Sci. B* **2015**, *16*, 709–719. [CrossRef]
111. Ouwerkerk, D.; Maguire, A.; McMillen, L.; Klieve, A.V. Hydrogen utilising bacteria from the forestomach of eastern grey (*Macropus giganteus*) and red (*Macropus rufus*) kangaroos. *Anim. Prod. Sci.* **2009**, *49*, 1043–1051. [CrossRef]
112. Gagen, E.J.; Wang, J.; Padmanabha, J.; Liu, J.; de Carvalho, I.P.C.; Liu, J.; Webb, R.; Al Jassim, R.; Morrison, M.; E Denman, S.; et al. Investigation of a new acetogen isolated from an enrichment of the tammar wallaby forestomach. *BMC Microbiol.* **2014**, *14*, 314. [CrossRef]
113. Breznak, J.A. Acetogenesis from carbon dioxide in termite guts. In *Acetogenesis*; Drake, H.L., Ed.; Chapman and Hall: New York, NY, USA, 1994; pp. 303–330.
114. Klieve, A.V.; Joblin, K.N. Comparison in hydrogen utilization of ruminal and marsupial reductive acetogens. In Proceedings of the PGGRC 5-Year Science Progress Report, 3rd International Greenhouse Gases and Animal Agriculture Conference, Christchurch, New Zealand, 26–29 November 2007; pp. 34–35.
115. Short, S.B.; Edwards, W.C. Sulfur (hydrogen-sulfide) toxicosis in cattle. *Vet. Hum. Toxicol.* **1989**, *31*, 451–453. [PubMed]
116. Odom, J.M.; Singleton, R., Jr. *The Sulfate-Reducing Bacteria: Contemporary Perspectives*; Springer-Verlag Inc.: New York, NY, USA, 1993; p. 289. [CrossRef]

117. Thomas, W.E.; Loosli, J.K.; Williams, H.H.; Maynard, L.A. The utilization of inorganic sulfates and urea nitrogen by lambs. *J. Nutr.* **1951**, *43*, 515–523. [CrossRef] [PubMed]
118. Huisingh, J.; McNeill, J.J.; Matrone, G. Sulphate reduction by a *Desulfovibrio* species isolated from sheep rumen. *Appl. Microbiol.* **1974**, *28*, 489–497. [CrossRef] [PubMed]
119. Cummings, B.A.; Caldwell, D.R.; Gould, D.H.; Hamar, D.W. Identity and interactions of rumen microbes associated with dietary sulfate-induced polioencephalomalacia. *Am. J. Vet. Res.* **1995**, *56*, 1384–1389. [PubMed]
120. Kushkevych, I.; Kotrsová, V.; Dordević, D.; Buňková, L.; Vítězová, M.; Amedei, A. Hydrogen sulfide effects on the survival of lactobacilli with emphasis on the development of inflammatory bowel diseases. *Biomolecules* **2019**, *9*, 752. [CrossRef] [PubMed]
121. Ran, S.; Mu, C.; Zhu, W. Diversity and community pattern of sulfate-reducing bacteria in piglet gut. *J. Anim. Sci. Biotechnol.* **2019**, *10*, 40. [CrossRef]
122. McCann, J.C.; Segers, J.R.; Derakhshani, H.; Felix, T.L.; Khafipour, E.; Shike, D.W. Increasing corn distillers solubles alters the liquid fraction of the ruminal microbiome. *J. Anim. Sci.* **2017**, *95*, 3540–3551. [CrossRef]
123. Howard, B.H.; Hungate, R.E. *Desulfovibrio* of the sheep rumen. *Appl. Environ. Microbiol.* **1976**, *32*, 598–602. [CrossRef]
124. Paul, S.S.; Deb, S.M.; Singh, D. Isolation and characterization of novel sulphate-reducing *Fusobacterium* sp. and their effects on in vitro methane emission and digestion of wheat straw by rumen fluid from Indian riverine buffaloes. *Anim. Feed Sci. Technol.* **2011**, *166*, 132–140. [CrossRef]
125. NRC. *Nutrient Requirements of Sheep*, 6th ed.; National Academy Press: Washington, DC, USA, 1985.
126. NRC. *Nutrient Requirements of Beef Cattle*, 7th ed.; National Academy Press: Washington, DC, USA, 2000.
127. Whanger, P.D.; Matrone, G. Effect of dietary sulfur upon the production and absorption of lactate in sheep. *Biochim. Biophys. Acta* **1996**, *124*, 273–279. [CrossRef]
128. Rumsey, T.S. Effects of dietary sulfur addition and Synovex-S ear implants on feedlot steers fed an all-concentrate finishing diet. *J. Anim. Sci.* **1978**, *46*, 463–477. [CrossRef]
129. Whanger, P.D.; Matrone, G. Effect of dietary sulfur upon the fatty acid production in the rumen. *Biochim. Biophys. Acta* **1965**, *98*, 454–461. [CrossRef] [PubMed]
130. Hegarty, R.S.; Nolan, J.V.; Leng, R.A. The effects of protozoa and of supplementation with nitrogen and sulfur on digestion and microbial metabolism in the rumen of sheep. *Aust. J. Agric. Res.* **1994**, *45*, 1215–1227. [CrossRef]
131. Patterson, J.A.; Kung, L. Metabolism of DL-methionine and methionine analogs by rumen micro-organisms. *J. Dairy Sci.* **1988**, *71*, 3292–3301. [CrossRef] [PubMed]
132. Inthapanya, S.; Preston, T.R.; Khang, D.N.; Leng, R.A. Effect of potassium nitrate and urea as fermentable nitrogen sources on growth performance and methane emissions in local "Yellow" cattle fed lime ($Ca(OH)_2$) treated rice straw supplemented with fresh cassava foliage. *Livest. Res. Rural Dev.* **2012**, *24*, 27. Available online: http://www.lrrd.org/lrrd24/2/sang24027.htm (accessed on 25 October 2022).
133. Sophal, C.; Khang, D.N.; Preston, T.R.; Leng, R.A. Nitrate replacing urea as a fermentable N source decreases enteric methane production and increases the efficiency of feed utilization in Yellow cattle. *Livest. Res. Rural Dev.* **2013**, *25*, 113. Available online: http://www.lrrd.org/lrrd25/7/soph25113.htm (accessed on 25 October 2022).
134. Ascensão, A.M.D. Effects of Nitrate and Additional Effect of Probiotic on Methane Emissions and Dry Matter Intake in Nellore Bulls. Master's Thesis, Universidade de Trás-os-Montes e Alto Douro Departamento de Zootecnia, Vila Real, Portugal, 2010.
135. Wood, T.A.; Wallace, R.J.; Rowe, A.; Price, J.; Yáñez-Ruiz, D.R.; Murray, P.; Newbold, C.J. Encapsulated fumaric acid as a feed ingredient to decrease ruminal methane emissions. *Anim. Feed Sci. Technol.* **2009**, *152*, 62–71. [CrossRef]
136. Demeyer, D.I.; Henderickx, H.K. Competitive inhibition of in vitro methane production by mixed rumen bacteria. *Arch. Internat. Physiol.* **1967**, *75*, 157–159.
137. Kolver, E.S.; Aspin, P.W.; Jarvis, G.N.; Elborough, K.M.; Roche, J.R. Fumarate reduces methane production from pasture fermented in continuous culture. *Proc. N. Z. Soc. Anim.* **2004**, *64*, 155–159.
138. Newbold, C.J.; Ouda, J.O.; Lopez, S.; Nelson, N.; Omed, H.; Wallace, R.J.; Moss, A.R. Propionate precursors as possible alternative electron acceptors to methane in ruminal fermentation. In *Greenhouse Gases and Animal Agriculture*; Takahashi, J., Youngeds, B.A., Eds.; Elsevier: Amsterdam, The Netherlands, 2002; pp. 151–154.
139. Iwamoto, M.; Asanuma, N.; Hino, T. Effects of nitrate combined with fumarate on methanogenesis, fermentation, and cellulose digestion by rumen microbes in vitro. *Anim. Sci. J.* **1999**, *70*, 471–478. [CrossRef]
140. Asanuma, N.; Iwamoto, M.; Hino, T. Effect of the addition of fumarate on methane production by ruminal microorganisms in vitro. *J. Dairy Sci.* **1999**, *82*, 780–787. [CrossRef] [PubMed]
141. Arif, M.; Sarwar, M.; Mehr-un-Nisa; Hayat, Z.; Younas, M. Effect of supplementary sodium nitrate and sulphur on methane production and growth rates in sheep and goats fed forage-based diet low in true protein. *J. Anim. Plant Sci.* **2016**, *26*, 69–78.
142. Sarturi, J.O.; Erickson, G.E.; Klopfenstein, T.J.; Rolfe, K.M.; Buckner, C.D.; Luebbe, M.K. Impact of source of sulfur on ruminal hydrogen sulfide and logic for the ruminal available sulfur for reduction concept. *J. Anim. Sci.* **2013**, *91*, 3352–3359. [CrossRef]
143. Bray, A.C. Sulphur metabolism in sheep. II. Absorption of inorganic sulphate and inorganic sulphide from sheeps rumen. *Aust. J. Agric. Res.* **1969**, *20*, 739–748. [CrossRef]
144. Richter, E.L.; Drewnoski, M.E.; Hansen, S.L. Effect of increased dietary sulfur on beef steer mineral status, performance, and meat fatty acid composition. *J. Anim. Sci.* **2012**, *90*, 3945–3953. [CrossRef] [PubMed]

145. Drewnoski, M.E.; Doane, P.; Hansen, S.L. Ferric citrate decreases ruminal hydrogen sulphide concentrations in feedlot cattle fed diets high in sulphate. *Br. J. Nutr.* **2014**, *111*, 261–269. [CrossRef]
146. DiLorenzo, N.; Dahlen, C.R.; Diez-Gonzalez, F.; Lamb, G.C.; Larson, J.E.; DiCostanza, A. Effects of feeding polyclonal antibody preparations on rumen fermentation patterns, performance, and carcass characteristics of feedlot steers. *J. Anim. Sci.* **2008**, *86*, 3023–3032. [CrossRef] [PubMed]
147. Latham, E.A.; Anderson, R.C.; Pinchak, W.E.; Nisbet, D.J. Insights on Alterations to the rumen ecosystem by nitrate and nitrocompounds. *Front. Microbiol.* **2016**, *7*, 228. [CrossRef]
148. Payne, W.J. Reduction of nitrogenous oxides by microorganisms. *Bacteriol. Rev.* **1973**, *37*, 409–452. [CrossRef]
149. Baar, C.; Eppinger, M.; Raddatz, G.; Simon, J.; Lanz, C.; Klimmek, O.; Nandakumar, R.; Gross, R.; Rosinus, A.; Keller, H.; et al. Complete genome sequence and analysis of *Wolinella succinogenes*. *Proc. Natl. Acad. Sci. USA* **2003**, *100*, 11690–11695. [CrossRef]
150. Guo, W.S.; Schaefer, D.M.; Guo, X.X.; Ren, L.P.; Meng, Q.X. Use of nitrate-nitrogen as a sole dietary nitrogen source to inhibit ruminal methanogenesis and to improve microbial nitrogen synthesis in vitro. *Asian-Australas. J. Anim. Sci.* **2009**, *22*, 542–549. [CrossRef]
151. Brunet, R.C.; Garcia-Gil, L.J. Sulfide-induced dissimilatory nitrate reduction to ammonia in anaerobic freshwater sediments. *FEMS Microbiol. Ecol.* **1996**, *21*, 131–138. [CrossRef]
152. Leng, R.A.; Nolan, J.V. Nitrogen metabolism in the rumen. *J. Dairy Sci.* **1984**, *67*, 1072–1089. [CrossRef] [PubMed]
153. Cockburn, A.; Brambilla, G.; Fernandez, M.-L.; Arcella, D.; Bordajandi, L.R.; Cottrill, B.; van Peteghem, C.; Dorne, J.-L. Nitrite in feed: From animal health to human health. *Toxicol. Appl. Pharmacol.* **2013**, *270*, 209–217. [CrossRef] [PubMed]
154. Alaboudi, A.; Jones, G. Effect of acclimation to high nitrate intakes on some rumen fermentation parameters in sheep. *Can. J. Anim. Sci.* **1985**, *65*, 841–849. [CrossRef]
155. Hao, T.P.; Quang, D.H.; Preston, T.R.; Leng, R.A. Nitrate as a fermentable nitrogen supplement for goats fed forage based diets low in true protein. *Livest. Res. Rural Dev.* **2009**, *21*. Available online: https://www.lrrd.org/lrrd21/1/trin21010.htm (accessed on 25 October 2022).
156. Jones, G.A. Dissimilatory metabolism of nitrate by the rumen microbiota. *Can. J. Microbiol.* **1972**, *18*, 1783–1787. [CrossRef] [PubMed]
157. Sar, C.; Mwenya, B.; Pen, B.; Takaura, K.; Morikawa, R.; Tsujimoto, A.; Kuwaki, K.; Isogai, N.; Shinzato, I.; Asakura, Y.; et al. Effect of ruminal administration of *Escherichia coli* wild type or a genetically modified strain with enhanced high nitrite reductase activity on methane emission and nitrate toxicity in nitrate-infused sheep. *Br. J. Nutr.* **2005**, *94*, 691–697. [CrossRef]
158. Stewart, C.S.; Bryant, M.P. The rumen bacteria. In *The Rumen Microbial Ecosystem*; Hobson, P.N., Stewart, C.S., Eds.; Blackie Academic and Professional: London, UK, 1988; pp. 21–75.
159. Sar, C.; Mwenya, B.; Santoso, B.; Takaura, K.; Morikawa, R.; Isogai, N.; Asakura, Y.; Toride, Y.; Takahashi, J. Effect of *Escherichia coli* wild type or its derivative with high nitrite reductase activity on in vitro ruminal methanogenesis and nitrate/nitrite reduction. *J. Anim. Sci.* **2005**, *83*, 644–652. [CrossRef]
160. Sar, C.; Mwenya, B.; Santoso, B.; Takaura, K.; Morikawa, R.; Isogai, N.; Asakura, Y.; Toride, Y.; Takahashi, J. Effect of *Escherichia coli* W3110 on ruminal methanogenesis and nitrate/ nitrite reduction in vitro. *Anim. Feed Sci. Technol.* **2005**, *118*, 295–306. [CrossRef]
161. Mwenya, B.; Zhou, X.; Santoso, B.; Sar, C.; Gamo, Y.; Kobayashi, T.; Takahashi, J. Effects of probiotic-vitacogen and β 1-4 galacto-oligosaccharides supplementation on methanogenesis and energy and nitrogen utilization in dairy cows. *Asian-Australas. J. Anim. Sci.* **2004**, *17*, 349–354. [CrossRef]
162. Yoshida, J.; Nakamura, Y.; Nakamura, R. Effects of protozoal fraction and lactate on nitrate metabolism of microorganisms in sheep rumen. *Jpn. J. Zootech. Sci.* **1982**, *53*, 677–685. [CrossRef]
163. Lin, M.; Schaefer, D.M.; Guo, W.S.; Ren, L.P.; Meng, Q.X. Comparisons of in vitro nitrate reduction, methanogenesis, and fermentation acid profile among rumen bacterial, protozoal and fungal fractions. *Asian-Australas. J. Anim. Sci.* **2011**, *24*, 471–478. [CrossRef]
164. Iwamoto, M.; Asanuma, N.; Hino, T. Effects of energy substrates on nitrate reduction and nitrate reductase activity in a ruminal bacterium, *Selenomonas ruminantium*. *Anaerobe* **2001**, *7*, 315–321. [CrossRef]
165. Sophea, I.v.; Preston, T.R. Effect of different levels of supplementary potassium nitrate replacing urea on growth rates and methane production in goats fed rice straw, mimosa foliage and water spinach. *Livest. Res. Rural Dev.* **2011**, *23*, 71. Available online: http://www.lrrd.org/lrrd23/4/soph23071.htm (accessed on 25 October 2022).
166. Li, L.; Davis, J.; Nolan, J.; Hegarty, R. An initial investigation on rumen fermentation pattern and methane emission of sheep offered diets containing urea or nitrate as the nitrogen source. *Anim. Prod. Sci.* **2012**, *52*, 653–658. [CrossRef]
167. Zhou, Z.; Meng, Q.; Yu, Z. Effects of methanogenic inhibitors on methane production and abundances of methanogens and cellulolytic bacteria in in vitro ruminal cultures. *Appl. Environ. Microbiol.* **2011**, *77*, 2634–2639. [CrossRef]
168. Ungerfeld, E.M. Shifts in metabolic hydrogen sinks in the methanogenesis-inhibited ruminal fermentation: A meta-analysis. *Front. Microbiol.* **2015**, *6*, 37. [CrossRef] [PubMed]
169. Wang, K.; Nan, X.; Chu, K.; Tong, J.; Yang, L.; Zheng, S.; Zhao, G.; Jiang, L.; Xiong, B. Shifts of hydrogen metabolism from methanogenesis to propionate production in response to replacement of forage fiber with non-forage fiber sources in diets in vitro. *Front. Microbiol.* **2018**, *9*, 2764. [CrossRef]
170. Castillo, C.; Benedito, J.L.; Mendez, J.; Pereira, V.; Lopez-Alonso, M.; Miranda, M.; Hernández, J. Organic acids as a substitute for monensin in diets for beef cattle. *Anim. Feed Sci. Technol.* **2004**, *115*, 101–116. [CrossRef]

171. Baldwin, R.L.; Allison, J.M. Rumen metabolism. *J. Anim. Sci.* **1983**, *57*, 461–477. [PubMed]
172. Bergman, E.N. Energy contributions of volatile fatty acids from the gastrointestinal tract in various species. *Physiol. Rev.* **1990**, *70*, 567–590. [CrossRef] [PubMed]
173. Martin, S.A. Manipulation of ruminal fermentation with organic acids: A review. *J. Anim. Sci.* **1998**, *76*, 3123–3132. [CrossRef] [PubMed]
174. Nisbet, D.J.; Martin, S.A. Effects of fumarate, L-malate, and an *Aspergillus oryzae* fermentation extract on D-lactate utilization by the ruminal bacterium *Selenomonas ruminantium*. *Curr. Microbiol.* **1993**, *26*, 133–136. [CrossRef]
175. García-Martínez, R.; Ranilla, M.J.; Tejido, M.L.; Carro, M.D. Effects of disodium fumarate on in vitro rumen microbial growth, methane production and fermentation of diets differing in their forage: Concentrate ratio. *Br. J. Nutr.* **2005**, *94*, 71–77. [CrossRef]
176. Ungerfeld, E.M.; Kohn, R.A.; Wallace, R.J.; Newbold, C.J. A meta-analysis of fumarate effects on methane production in ruminal batch cultures. *J. Anim. Sci.* **2007**, *85*, 2556–2563. [CrossRef]
177. Castro-Montoya, J.; De Campeneere, S.; Van Ranst, G.; Fievez, V. Interactions between methane mitigation additives and basal substrates on in vitro methane and VFA production. *Anim. Feed Sci. Technol.* **2012**, *176*, 47–60. [CrossRef]
178. Wallace, R.J.; Wood, T.A.; Rowe, A.; Price, J.; Yanez, D.R.; Williams, S.P.; Newbold, C. Encapsulated fumaric acid as a means of decreasing ruminal methane emissions. *Int. Congr. Ser.* **2006**, *1293*, 148–151. [CrossRef]
179. Itabashi, H.; Bayaru, E.; Kanda, S.; Nishida, T.; Ando, S.; Ishida, M.; Itoh, T.; Isobe, Y.; Nagara, K.; Takei, K. Effect of Salinomycin (SL) or SL plus Fumaric Acid on rumen fermentation and methane production in cattle. *Asian-Australas. J. Anim. Sci.* **2000**, *13*, 287.
180. Beauchemin, K.A.; McGinn, S.M. Methane emissions from beef cattle: Effects of fumaric acid, essential oil, and canola oil. *J. Anim. Sci.* **2006**, *84*, 1489–1496. [CrossRef]
181. Martin, S.A.; Streeter, M.N. Effect of malate on in vitro mixed ruminal microorganism fermentation. *J. Anim. Sci.* **1995**, *73*, 2141–2145. [CrossRef]
182. Devant, M.; Bach, A.; García, J.A. Effect of malate supplementation to dairy cows on rumen fermentation and milk production in early lactation. *J. Appl. Anim. Res.* **2007**, *31*, 169–172. [CrossRef]
183. Kung, L., Jr.; Huber, J.T.; Krummerey, J.D.; Allison, L.; Cook, R.M. Influence of adding malic acid to dairy cattle rations on milk production, rumen volatile acids, digestibility, and nitrogen utilization. *J. Dairy Sci.* **1982**, *65*, 1170–1174. [CrossRef]
184. Sanson, D.W.; Stallcup, O.T. Growth response and serum constituents of Holstein bulls fed malic acid. *Nutr. Rep. Int.* **1984**, *30*, 1261–1267.
185. Callaway, T.R.; Martin, S.A. Effects of organic acid and monensin treatment on in vitro mixed ruminal microorganism fermentation of cracked corn. *J. Anim. Sci.* **1996**, *74*, 1982–1989. [CrossRef]
186. Carro, M.D.; López, S.; Valdés, C.; Ovejero, F.J. Effect of D, L-malate on mixed ruminal microorganism fermentation using the rumen simulation technique (RUSITEC). *Anim. Feed Sci. Technol.* **1999**, *79*, 279–288. [CrossRef]
187. Carro, M.D.; Ranilla, M.J. Effect of the addition of malate on in vitro rumen fermentation of cereal grains. *Br. J. Nutr.* **2003**, *89*, 181–188. [CrossRef] [PubMed]
188. Carro, M.D.; Ranilla, M.J. Influence of different concentrations of disodium fumarate on methane production and fermentation of concentrate feeds by rumen micro-organisms in vitro. *Br. J. Nutr.* **2003**, *90*, 617–623. [CrossRef]
189. Asanuma, N.; Hino, T. Activity and properties of fumarate reductase in ruminal bacteria. *J. Gen. Appl. Microbiol.* **2000**, *46*, 119–125. [CrossRef]
190. Mamuad, L.; Kim, S.H.; Jeong, C.D.; Choi, Y.J.; Jeon, C.O.; Lee, S.S. Effect of fumarate reducing bacteria on in vitro rumen fermentation, methane mitigation and microbial diversity. *J. Microbiol.* **2014**, *52*, 120–128. [CrossRef]
191. Astuti, W.D.; Wiryawan, K.G.; Wina, E.; Widyastuti, Y.; Suharti, S.; Ridwan, R. Effects of selected *Lactobacillus plantarum* as probiotic on in vitro ruminal fermentation and microbial population. *Pak. J. Nutr.* **2018**, *17*, 131–139. [CrossRef]
192. Knapp, J.R.; Laur, G.L.; Vadas, P.A.; Weiss, W.P.; Tricarico, J.M. Invited review: Enteric methane in dairy cattle production: Quantifying the opportunities and impact of reducing emissions. *J. Dairy Sci.* **2014**, *97*, 3231–3261. [CrossRef] [PubMed]
193. Kaewpila, C.; Gunun, P.; Kesorn, P.; Subepang, S.; Thip-uten, S.; Cai, Y.; Pholsen, S.; Cherdthong, A.; Khota, W. Improving ensiling characteristics by adding lactic acid bacteria modifies in vitro digestibility and methane production of forage-sorghum mixture silage. *Sci. Rep.* **2021**, *11*, 1968. [CrossRef] [PubMed]
194. Varnava, K.G.; Ronimus, R.S.; Sarojini, V. A review on comparative mechanistic studies of antimicrobial peptides against archaea. *Biotechnol. Bioeng.* **2017**, *114*, 2457–2473. [CrossRef]
195. Jeyanathan, J.; Martin, C.; Morgavi, D.P. Screening of bacterial direct-fed microbials for their antimethanogenic potential in vitro and assessment of their effect on ruminal fermentation and microbial profiles in sheep. *J. Anim. Sci.* **2016**, *94*, 739–750. [CrossRef]
196. Callaway, T.R.; Carneiro De Melo, A.M.S.; Russell, J.B. The effect of nisin and monensin on ruminal fermentations in vitro. *Curr. Microbiol.* **1997**, *35*, 90–96. [CrossRef]
197. Santoso, B.; Mwenya, B.; Sar, C.; Gamo, Y.; Kobayashi, T.; Morikawa, R.; Kimura, K.; Mizukoshi, H.; Takahashi, J. Effects of supplementing galacto-oligosaccharides, *Yucca schidigera* or nisin on ruminal methanogenesis, nitrogen and energy metabolism in sheep. *Livest. Prod. Sci.* **2004**, *91*, 209–217. [CrossRef]
198. Renuka; Puniya, M.; Sharma, A.; Malik, R.; Upadhyay, R.C.; Puniya, A.K. Influence of pediocin and enterocin on in-vitro methane, gas production and digestibility. *Int. J. Curr. Microbiol. Appl. Sci.* **2013**, *2*, 132–142.

199. Cao, Y.; Takahashi, T.; Horiguchi, K.; Yoshida, N. Effect of adding lactic acid bacteria and molasses on fermentation quality and in vitro ruminal digestion of total mixed ration silage prepared with whole crop rice. *Grassl. Sci.* **2010**, *56*, 19–25. [CrossRef]
200. Cao, Y.; Cai, Y.; Takahashi, T.; Yoshida, N.; Tohno, M.; Uegaki, R.; Nonaka, K.; Terada, F. Effect of lactic acid bacteria inoculant and beet pulp addition on fermentation characteristics and in vitro ruminal digestion of vegetable residue silage. *J. Dairy Sci.* **2011**, *94*, 3902–3912. [CrossRef]
201. Huyen, N.T.; Martinez, I.; Pellikaan, W. Using lactic acid bacteria as silage inoculants or direct-fed microbials to improve in vitro degradability and reduce methane emissions in dairy cows. *Agronomy* **2020**, *10*, 1482. [CrossRef]
202. Guo, G.C.Q.; Shen, C.; Liu, Q.; Zhang, S.; Shao, T.; Wang, C.; Wang, Y.; Xu, Q.; Huo, W. The effect of lactic acid bacteria inoculums on in vitro rumen fermentation, methane production, ruminal cellulolytic bacteria populations and cellulase activities of corn stover silage. *J. Integr. Agric.* **2020**, *19*, 838–847. [CrossRef]

methane

MDPI

Review

Genetic Improvement and Nutrigenomic Management of Ruminants to Achieve Enteric Methane Mitigation: A Review

Vasfiye Kader Esen [1], Valiollah Palangi [2,*] and Selim Esen [3]

[1] Sheep Breeding Research Institute, Balikesir 10200, Turkey
[2] Department of Animal Science, Agricultural Faculty, Ataturk University, Erzurum 25240, Turkey
[3] Balikesir Directorate of Provincial Agriculture and Forestry, Republic of Turkey Ministry of Agriculture and Forestry, Balikesir 10470, Turkey
* Correspondence: valiollah.palangi12@ogr.atauni.edu.tr

Abstract: A significant portion of global greenhouse gas emissions is attributed to methane (CH_4), the primary greenhouse gas released by dairy animals. Thus, livestock farming has a new challenge in reducing enteric CH_4 for sustainability. In anaerobic microbial ecosystems such as the rumen, carbohydrates are converted into short-chain, volatile fatty acids that animals use for energy and protein synthesis. It is, therefore, essential to understand rumen physiology, population dynamics, and diversity to target methanogens. Thus far, numerous CH_4 mitigation strategies have been studied, including feeding management, nutrition, rumen modification, genetics, and other approaches for increasing animal production. As new molecular techniques are developed, scientists have more opportunities to select animals with higher genetic merit through next-generation sequencing. The amount of CH_4 produced per unit of milk or meat can be permanently and cumulatively reduced through genetic selection. Developing eco-friendly and practical nutrigenomic approaches to mitigating CH_4 and increasing ruminant productivity is possible using next-generation sequencing techniques. Therefore, this review summarizes current genetic and nutrigenomic approaches to reducing enteric CH_4 production without posing any danger to animals or the environment.

Keywords: methane; mitigation; genetic; nutrigenomic

1. Introduction

In response to the increasing population rate and growing animal protein demand, the ruminant husbandry or livestock sector is experiencing rapid structural and functional changes. In particular, consumption of livestock products has increased twofold in the past few decades, while milk supply and demand have increased by 26% and 2.4% annually, respectively, and are expected to increase by 25% in the near future [1,2]. In addition to supporting approximately 1.3 billion producers and retailers, the livestock sector contributes between 40 and 50% of the agricultural gross domestic product [2]. It is necessary to underscore, however, that livestock, although of great social and economic importance, is also one of the largest sources of greenhouse gases (GHG) worldwide, particularly enteric methane (CH_4) [3]. Approximately 13% of the world's GHG emissions come from livestock farming; CH_4 emissions, in particular, are important contributors since their global warming potential is 28 times higher than carbon dioxide's [4]. As this carbon cannot be utilized as an energy source in ruminants, CH_4 production has a negative impact on ruminant efficiency and profitability, as well as on the environment [5].

Ruminants produce CH_4 as a natural reaction during the microbial fermentation of its feed in the rumen [6]. In cattle, for example, enteric CH_4 is mostly produced in the rumen (87–90%) and in the large intestine (10–13%) [7,8]. Various microbial species participate in the conversion of feed material to CH_4 in the rumen; methanogenic archaea perform the final step [9]. There has been significant evidence that genetic variation of the host animals impacts bacterial community composition and CH_4 production [10]. Furthermore, numerous studies have demonstrated that methane production is closely related to the weight of the animal, the amount of dry matter ingested, and the amount of gross energy ingested [11,12]. It is recognized that methane production is a hereditable trait, and that genetic selection for low-emitting ruminants is an effective mitigation option, assuming feed intake and animal productivity remain unchanged [13,14].

There is little information on how fermentative organisms interact with methanogens, but it is likely that other functional groups of microbes influence CH_4 production in rumens heavily. In addition to producing substrates necessary for the survival of methanogens, they also affect their numbers or other microbiota members responsible for producing methanogenic substrates [15–17]. To achieve a balance between food production and greenhouse gas emissions, microbiologists and nutritional scientists are investigating the genomes of the rumen microbiome to understand their function in terms of feed conversion efficiency, and plant cell wall degradation. Various strategies have been developed to mitigate enteric CH_4 emissions, such as alternate electron receptors [18], ionophore [19], enzymes [20], probiotic cultures [21], plant secondary metabolites [22]. Providing ruminants with forages that are more digestible or increasing their consumption of energy-dense feed is another effective CH_4 strategy [3]. Mitigating enteric CH_4 can also be achieved by improving pasture quality [1] and managing grazing [23], using antibiotics [24], vaccinations [25], or defaunation [26]. In any cases, it should be noted that developing strategies to reduce ruminant-derived CH_4 and alter microbiomes associated with ruminants will be one of the most significant challenges of the century and deserve a great deal of attention.

Despite the abundance of genetic improvement and nutrigenomic studies on CH_4 reduction, articles summarizing these studies are scarce. Hence, given the importance of the topic and the fast pace of growing knowledge in the area, in this article, the authors have tried to focus on genetic improvement and nutrigenomic management-based approaches to reduce enteric CH_4 emissions. In the current study, genetic strategies that can affect CH_4 production were classified into two major sections: 1–Manipulating ruminants via genetic selection and 2–Manipulation of the rumen microbiome.

2. Manipulating Ruminants via Genetic Selection

One of the methods, which can be classified in the manipulating the animal section, is genetically selecting ruminants to produce less CH_4. Genetic selection is an attractive solution, due to its cumulative and permanent nature; however, since selection is carried out over generations, it requires additive genetic variation and time to take effect [27]. Furthermore, genetic selection involves recording the CH_4 of a large number of animals, which is costly; therefore, to achieve a more accurate genetic evaluation, it is essential to use phenotypes that are precise and consistent [28].

In ruminants, approximately 98% of the CH_4 released is emitted directly from the rumen, absorbed from the rumen and hindgut into the blood, and exhaled from the lungs, while the remaining 2% is expelled as flatulence [27] and the rate of CH_4 emission varies with day [29], physiological state (growing, lactating and non-lactating) [30], and during lactation (early, mid, peak and late) [31], as well as between lactation periods [27]. In order to understand the implications of selection based on CH_4 emissions recorded at a particular point in time and to optimize selection strategies, it is essential to determine the relationship between CH_4 emissions recorded at various points during an animal's life and its phenotypic and genetic characteristics [27].

Identifying genetic correlations between one trait and other traits plays an important role in evaluating the effectiveness of the current breeding strategy. In order to reduce CH_4 emissions, we might select high-yielding animals for two main reasons. The first reason is that they are more productive, convert feed more efficiently, and require less maintenance. The second reason is that fewer animals with better productivity will require fewer animals to achieve a target production level [1]. The advantage of selecting high-yielding animals is that it allows cumulative and permanent changes; however, it depends on additive genetic variation and time to be effective since selection takes generations [27].

A repeatable and highly correlated method should be used with the existing traits in the selection index. Many researchers have speculated that a novel method of measuring the concentrations of CH_4 released by dairy cows during milking may provide valuable information regarding the daily emissions of individual dairy cows [32,33]. Moreover, the measurement of enteric CH_4 emissions in dairy cows is more straightforward than in cattle from another production system [33]. Typically, dairy cows are milked 3 times daily by standard milking parlors and 6 times daily by automatic milking systems [32]. Consequently, individual enteric CH_4 emissions can be monitored on an ongoing basis without the need for invasive techniques [32]. Thus, a number of researchers recommend the integration of measuring systems with automatic milking systems to be able to provide accurate, repeatable information on the amount of CH_4 emitted by dairy cows [32,34]. A number of techniques and devices have been used for this purpose, including respiration chambers [35], portable accumulation chambers [36], hexafluoride tracer methods [23], laser CH_4 detectors [37], micrometeorological methods [38], and more recently, measurements of milk in the mid-infrared range [39]. A detailed representation of the methane production and methane yield results obtained using different methods is given in Table 1.

Table 1. Comparison of techniques used to measure methane emissions from ruminants.

Methods	Animal	Breed	Feed Intake	MeP	MeY	References
RC	Heifers	Hereford × Friesian	*ad libitum* (10.9–12.2 kg DM per day)	265	24.5	[40]
	Dairy cows	German Holstein (early lactation)	*ad libitum* TMR (grass silage, corn silage, barley straw, hay, concentrate, corn meal, canola seed meal, soybean meal, wheat, soybean oil)	346.4	22.2	[31]
	Beef steers	Brahman (*Bos indicus*) and Belmont Red (*Bos taurus x African Sanga*)	Rhodes grass pasture (*Chloris gayana*) grazed	114.3	30.1	[38]
	Lambs	Coopworth, Romney, Perendale, Texel, and composite breeds	*ad libitum* pasture allowance	24	16	[36]
PAC	Lambs	Coopworth, Romney, Perendale, Texel, and composite breeds	*ad libitum* pasture allowance	7.5	-	[36]
SF_6	Heifers	Hereford × Friesian	*ad libitum* (10.9–12.2 kg DM per day)	272	22.4	[40]
	Cows	Angus	fed with 88% DM, 14% CP, 67% DM digestibility, and ME content of 9 MJ/kg DM	132.6	21.9	[41]
	Cows	Australian Holstein	fed a diet based on alfalfa supplemented with around 6 kg of crushed wheat per day	110.5	17.5	[41]
MHC	Heifers	Hereford × Friesian	*ad libitum* (10.9–12.2 kg DM per day)	323	26.8	[40]
	Dairy cows	German Holstein (early lactation)	*ad libitum* TMR (grass silage, corn silage, barley straw, hay, concentrate, corn meal, canola seed meal, soybean meal, wheat, soybean oil)	338.2	20.1	[31]
MM	Beef steers	Brahman (*Bos indicus*) and Belmont Red (*Bos taurus x African Sanga*)	Rhodes grass pasture (*Chloris gayana*) grazed	136.1	29.7	[38]
MIR	Dairy cows	Spanish Holstein	-	182.5	38.0	[42]

MeP: Methane production (g/d); MeY: methane yield (g/kg DMI (dry matter intake)); RC: respiration chambers; PAC: portable accumulation chambers; SF_6: hexafluoride tracer methods; MHC: mobile head-chamber (also known as GreenFeed); MM: micrometeorological methods; MIR: mid-infrared range method; TMR: total mixed ration; DM: dry matter; CP: crude protein; ME: metabolizable energy.

An accurate and preferably inexpensive phenotype is essential for making genetic evaluations relevant to the traits of interest. To genetically reduce GHG emissions, methane production (MeP, g/d), methane intensity (MI, g/kg at weight at test day (WT, kg)), methane yield (MeY, g/kg dry matter intake (DMI, kg/d)), and residual methane production (RMeP, g/d) are the most preferred phenotypes [28]. However, it might be challenging to properly incorporate MI, MeY, and RMeP into the selection index and explain to breeders why these phenotypes are the least frequently used. Therefore, the most correct way might be to use MeP in conjunction with its correlation structure with the milk yield, milk contents, live weight, and feed intake in the selection index [28]. Meanwhile, it should be noted that various factors may affect the production of CH_4 in individual cows, including their feed intake, dry matter content, feed composition, as well as fermentation rate in the rumen [43,44].

In contrast to MeP, several researchers have pointed out that RMeP is adjusted for traits that influence CH_4 outputs [28,45]. RMeP is computed as the difference between actual and predicted CH_4 outputs based on a subset of measured phenotypes [41]. As a statistical tool, RMeP offers advantages because its relationship with phenotypes used in its calculation is generally uncorrelated and it accurately predicts selection response [45]. Furthermore, Richardson et al. [45] reported that genetic interactions may still exist even after correcting RMeP traits for influential traits. A significant negative genetic correlation, for example, was found between the dry matter intake (DMI)-MeY (−0.60) and the energy-corrected milk (ECM)-MI (−0.73). A detailed comparison of genetic and phenotypic correlations between methane emission traits in ruminants is also provided in Table 2.

Table 2. Comparision of genetic and phenotypic correlations between methane emission traits in ruminants.

Methods	Animal	Breed	GC	PC	References
RC	Lambs	Coopworth, Romney, Perendale, Texel, and composite breeds	BW-MeP: 0.83 BW-MeY: 0.02	BW-MeP: 0.61 BW-MeY: 0.003	[36]
RC	Cattle	Angus	-	MeP-DMI: 0.65, MeY-DMI: −0.02	[35]
RC	Dairy cows	Holstein and Jersey [1]	MeP-DMI: 0.70, BW-MeP:0.54, ECM-MeP:0.66	-	[46]
RC	Dairy cows	Holstein and Jersey [2]	MeP-DMI: 0.49, BW-MeP:0.24, ECM-MeP:0.52	-	[46]
PAC	Lambs	Coopworth, Romney, Perendale, Texel, and composite breeds	BW-MeP: 0.59	BW-MeP: 0.31	[36]
MIR	Dairy cows	Spanish Holstein	MeP-RT: −0.43, MeP-MY: 0.21, MeP-PY: 0.31, MeP-FY: 0.29	MeP-MY: 0.05, MeP-PY: −0.03, MeP-FY: 0.00	[4,42]
SF_6	Cows	Holstein and Angus	MeP: DMI: 0.83, MeP-BW: 0.80, MeY-DMI: 0.08, MeY-BW: 0.05	MeP: DMI: 0.70, MeP-BW: 0.67, MeY-DMI: −0.00, MeY-BW: 0.04	[41]
SF_6	Dairy cows	-	MeP-DMI: 0.42, MEY-DMI: −0.60	MeP-DMI: 0.49, MEY-DMI: −0.27	[45]
Calculation	Heifers	Holstein-Friesian	-	MeP-FPCM: 0.26, MeP-DMI:0.99, MeP-RFI: 0.72	[9]

[1]: based on individual-level correlations; [2]: based on phenotypic level correlations; GC: genotypic correlation; PC: Phenotypic correlation; RC: respiration chambers; PAC: portable accumulation chambers; SF_6: hexafluoride tracer methods; MIR: mid-infrared range method; MeP: Methane production (g/d); MeY: methane yield (g/kg DMI (dry matter intake)); RT: rumination time (min/d); MY: milk yield (kg/d); PY: protein yield (kg/d); FY: fat yield (kg/d); FPCM: fat- and protein-corrected milk production (kg/d); RFI: residual feed intake (MJ/d).

A decrease in MeY has been observed with increasing DMI, even with small increases [36], while an increase in CH_4/CO_2 ratio was observed with increasing DMI [40]. Furthermore, Brask et al. [47] found that lactating dairy cows' CH_4/CO_2 ratio is higher at night than during the day, while it decreases with increasing time off pasture in sheep [48]. According to Robinson et al. [48], sheep's CH_4 emissions declined dramatically with increasing time away from pasture, while their CO_2 emissions declined only slightly.

Based on methane production rates, Kitelman et al. [49] identified three distinct ruminotypes in sheep: ruminotype Q (low-CH_4 production; higher abundances of propionate-producing e.g., *Quinella ovalis*), ruminotype S (low-CH_4 production; higher abundances of lactate- and succinate-producing bacteria, e.g., *Fibrobacter* spp., *Kandleria vitulina, Olsenella* spp., *Prevotella bryantii*, and *Sharpea azabuensis*), or ruminotype H (high-CH_4 production; higher abundances of *Ruminococcaceae, Lachnospiraceae, Catabacteriaceae, Coprococcus, Clostridiales, Prevotella, Bacteroidales*, and *Alphaproteobacteria*). Previous research also indicated that sheep selected for lower enteric CH_4 emissions have smaller rumens [50], a faster outflow of rumen liquor, and more VFA absorption gene expression in the rumen wall [51], which could lead to a reduction of ruminal VFA concentration. The muscle genes were identified as the best candidates as causal genes affecting CH_4 yield by [52]. In their study of the correlation between genetic parameters and CH_4 production, Jonker et al. [36] found that selectability for reduced CH_4 production can also lead to the selection of animals with a lower CH_4 yield due to dry matter intake [53] or animal genetics. It was found that CH_4/DMI did not correlate clearly with fecal matter output [54]. According to Renand et al. [54], fecal matter output and retention have a weak correlation, while fecal output has a moderate correlation with CH_4/DMI. Moreover, CH_4 production is estimated to have a low genetic correlation with remaining traits [5].

Using genetic selection as a tool can reduce CH_4 emissions and improve ruminant energy efficiency simultaneously without negatively affecting their important economic traits [45]. Breeding objectives for ruminants should include reducing CH_4 emissions while preserving economic traits that are important to sustainability.

3. Manipulation of the Rumen Microbiome

3.1. Rumen Microbiome

Recently, efforts have been directed toward characterizing the rumen microbiome and its function in order to implement nutritional and selective breeding strategies to alter it. Both the host and the ruminal microbiota affect livestock traits such as efficiency and sustainability, including CH_4 production, and are partly controlled by the host genotype [55]. The rumen microbiota, however, is highly dependent on the ruminant species, diet, and geographic location, leading to different rumen microbiome profiles and dietary nutrient utilization in ruminants from tropical and temperate environments [56]. Although these factors are present, microbial communities are generally stable due to ecological redundancy and resilience to external and internal perturbations [7].

In newborn ruminants, gastrointestinal tracts (GITs) are considered sterile upon birth, but methanogens and fibrolytic bacteria appear within 20 min of birth [57]. A day after parturition, cellulolytic bacteria, such as *Ruminococcus flavefaciens, Ruminococcus albus, Prevotella* species are predominant in the rumen microbiota [58], whereas xylanase and amylase (carbohydrate degrading enzymes) activity, as well as VFAs production, are evident within 2 days [59]. Furthermore, anaerobic fungi and methanogens begin to colonize the neonatal rumen around 8 to 10 days postpartum, but protozoa do not appear until 15 days [24]. Among the microbiomes found in adult's GITs of ruminants, anaerobic bacteria are the most abundant (10^{10-11}/mL), followed by archaea (methanogens; 10^{8-9}/mL), protozoa (10^6/mL), and fungi (10^6/mL) that digest feed together [60]. Compared to mature animals, pre-ruminant methanogenic archaea produce more CH_4 from a wide range of substrates, possibly because the *Methanobacteriaceae* dominate the rumen around 3 months of age [61]. In later stages of life, changing the rumen's microbial ecology is more difficult [62]. Thus, several studies have shown that a favorable microbiome can be implanted into the rumen microbiome early in life through dietary or management intervention [24,63].

Several studies have indicated that the seven most abundant bacterial groups (*Prevotella, Butyrivibrio,* and *Ruminococcus, Lachnospiraceae (unclassified), Ruminococcaceae, Bacteroidales,* and *Clostridiales*) are found in many species of ruminants, accounting for 67.1% of all sequenced data, and are defined as "**core bacterial microbiomes**" at the genus or higher level [64,65]. Except for *Butyrivibrio*, none of these groups are adequately represented by characterized cultures, nor are their functions understood [66]. The largest clade of archaea, *Methanobrevibacter gottschalkii* and *Methanobrevibacter ruminantium*, represents 74% of all archaea. On the other hand, an archaeal community in the rumen was composed of five dominant methanogen groups, as well as one *Methanosphaera* spp. and two *Methanomassiliicoccaceae*-associated groups, accounting for 89.2%. This implies that rumen archaea have less diversity than rumen bacteria [65,67].

3.2. Manipulation of the Rumen Microbiome via Nutrigenomic Approaches

Despite the lack of clarity regarding the relationship between fermentative organisms and methanogens, there are a number of functional groups of microbes that may have a significant impact on CH_4 production in the rumen, either by producing substrates that promote methanogen survival or by altering the number of methanogens or other microbes that produce methanogenic substrates [15]. Therefore, changing the dynamics of rumen fermentation by modifying other microbial groups may be an effective method for reducing CH_4 production.

Rumen hydrogen production is directly influenced by the pattern of fermentation of volatile fatty acids (VFAs), in other words ruminal VFA profile is an indicator of microbial activity [68]. As a result of glycolysis and final synthesis of VFA in the rumen, hydrogen is produced. For instance, 1 mol of acetate yields 2 mol of hydrogen, whereas 1 mol of propionate yields only 1 mol of hydrogen [69]. It is possible, therefore, to reduce CH_4 emissions via interventions affecting the rumen microbiome's acetate/propionate ratio [70], increasing bacteria species that compete for hydrogen (such as sulfate reducers and acetogens) [71], or inhibiting protozoa, which can cause hydrogen production to be reduced [72].

As an ionophore (which have antibiotic capacity), monensin is used to prevent ketosis in dairy cows and to increase production [73]. It alters the rumen microbiota, leading to increased hepatic gluconeogenesis and, therefore, an increase in the animal's energy supply when added to the diet, which results in an increase in propionate production [74]. In order to determine the differential effects of monensin and a mixture of essential oils on rumen microbiota composition in transition dairy cows, an experiment was conducted. Monensin decreased the relative abundance of 23 OTUs from the phyla *Bacteroidetes* and *Firmicutes*, whereas it increased the abundance of 10 OTUs from the phyla *Actinobacteria, Bacteroidetes, Cyanobacteria,* and *Firmicutes* [72]. As a result of monensin supplementation, the *Butyrivibrio* genus (which produces butyrate and acetate) was inhibited and propionate production was increased, whereas *Prevotella* and *Ruminococcaceae* (succinate and propionate producers, respectively) increased in abundance due to the rapid conversion of succinate to propionate by succinate-decarboxylating bacteria in the rumen [72]. It has previously been demonstrated that *Prevotella* species compete with methanogens for hydrogen use, rerouting it to propionic acid production and reducing methanogen availability [75].

A reserve polysaccharide in plants, inulin influences the proliferation of probiotics, such as *Lactobacillus* and *Bifidobacteria*, and inhibits bacterial pathogens, such as *Clostridium pneumonia, Enterococcus* and mold [76]. The effects of inulin on goat rumen fermentation and microbial growth were investigated by Zhao et al. [77] using rumen simulation technology. It was determined that inulin treatment decreased acetate concentrations, acetate ratios, and CH_4 production, while it increased butyrate concentrations. Furthermore, inulin inhibited *Fibrobacter succinogenes* and *Ruminococcus flavefaciens* growth, indicating that it might suppress the growth of rumen bacteria that decompose cellulose. In a study of finishing Simmental × Luxi crossbred beef steers fed a high or low-concentrate diet supplemented with 2% inulin (*wt/wt*), Tian et al. [78] demonstrated similar results regarding rumen fermentation and bacterial microbiota. The addition of inulin to a finishing beef diet resulted in a shift in the fermentation of acetate to propionate and butyrate, resulting in an increase in the α-diversity indexes of rumen bacteria, particularly *Bacteroides* and *Firmicutes*, which were significantly more abundant [78]. Furthermore previous reports has also demonstrated that cellulolytic *Ruminococcus* species (*Ruminococcus flavefaciens* and *Ruminococcus albus*)—along with *Butyrivibrio fibrisolvens*—were unable to grow in medium containing long-chain polyunsaturated fatty acids (such as docosahexaenoic acid or eicosapentaenoic acid) [79,80]. Additionally, Burdick et al. [81] found that *Methanobrevibacter gottschalkii* relative abundance increased while *Methanosphaera* spp. ISO3-F5 relative abundance decreased, but the ratio *Methanobrevibacter gottschalkii*: *Methanobrevibacter ruminantium*, associated with lower CH_4 emissions [82], did not change in lactating Holstein cows supplemented with medium-chain fatty acids.

During the course of researching 12 Nordic macroalgae species for anti-methanogenic properties, Pandey et al. [83] found that polyphenol-rich brown species, such as *Fucus vesiculosus* and *Ascophyllum nodosum*, significantly reduced feed degradability due to suppressed cellulolytic bacteria (*Ruminococcus* spp., *Lacnospiraceae* spp., *Rikenellaceae* RC9). These two macroalgae have been shown to reduce CH_4 production by 62.6 and 48.2%, respectively, and to reduce methanogenic archaea in rumens (such as *Methanobrevibacter* spp.). On the other hand, the reduction was not directly correlated with polyphenol concentrations overall.

Aside from converting carbohydrates into succinate and acetate, the *Succinivibrionaceae* family (*Succinivibrio*, *Ruminobacter*, *Anaerobiospirillum*, and *Succinimonas*) also produces hydrogen and acetate, which reduces CH_4 emissions [84]. Additionally, previous research suggests that feed efficiency of beef cattle [85] and milk protein of dairy cows [86] are highly related to the *Succinivibrio* spp. population in the rumen due to its function of producing succinate, the precursor of propionate.

Dairy cow diets are often supplemented with lipids to increase energy content; however, such lipid sources can change the composition of rumen bacteria and may affect biohydrogenation processes. For instance, Vargas-Bello-Perez et al. [87] investigated the effects of two dietary lipids on bacterial populations and fatty acid profiles in non-lactating Holstein cows' rumen digesta by utilizing soybean oil (an unsaturated oil source) and hydrogenated palm oil (a saturated oil source) and they found that dietary treatments had no effect on *Fibrobacter succinogenes*, *Butyrivibrio proteoclasticus*, and *Anaerovibrio lipolytica* loads. The same authors noted that with hydrogenated palm oil supplementation, the load of *Prevotella bryantii* significantly increased compared with control. In order to develop effective CH_4 mitigation strategies, Gruninger et al. [88] added 3-nitrooxypropanol (3-NOP), canola oil, and their combination to a high-forage diet (90% barley silage) of beef cattle. The authors noted that 3-NOP decreased the abundance of *Methanobrevibacter* and increased the abundance of *Bacteroidetes*, whereas canola oil significantly reduced the abundance of protozoa and fiber-degrading microbes in rumens, but did not significantly alter the abundance of rumen methanogens [88]. On the other hand, soybean or linseed oil (4% of DM) appeared to be more effective at decreasing *Butyrivibrio fibrisolvens*, *Ruminococcus albus*, and *Fibrobacter succinogenes* by 18, 42, and 67%, respectively [89]. An increase in the relative abundance of *Prevotella* and *Dialister* bacteria was also observed following the addition of oregano essential oil, which indicates that it has the potential to manipulate ruminal fermentation and reduce CH_4 emissions [90].

Research has also shown that fiber digestion and CH_4 production are related, particularly in the rumen where fibrolytic bacteria are involved in H_2 production (e.g., *Ruminococcus, Eubacterium*) and consumption (e.g., *Bacteroidetes*) [91]. Acetic acid production in the rumen is also closely linked with the production of H_2, which increases the substrate availability for CH_4 production [92]. A reduction in fiber content in diet is known to decrease *Bacteroides* abundance, as Pitta et al. [93] demonstrated. In addition, Mu et al. [94] reported that reducing fibrous substrate could be one explanation for *Fibrobacter* spp. (associated with cellulose degradation) and *Alistipes* spp. (associated with oligosaccharide degradation and butyric acid production [95]) declines in lactating Holstein cows fed high-grain diets. It is possible that *Fibrobacter succinogenes*'s cellulolytic activity stimulated *Prevotella ruminicola*'s production of propionic acid, since these bacteria consume only released sugars during cellulose digestion [95,96]. The rumen acetate proportion and the growth of *Ruminococcus albus* increased when forage concentrations were high in mixed diets [97]. Calves supplemented with *Prevotella* and cellulolytic bacteria (*Ruminococcus flavefaciens* and *Ruminococcus albus*) also showed similar results [98]. Furthermore, *Prevotella* may dominate the rumen in animals fed high-fiber diets, whereas *Bacteroidetes* may occupy a greater proportion in hay-supplemented animals, leading to increased rumen size [99,100].

4. Future Perspectives

In line with ruminal methane production reduction efforts, numerous methane mitigation strategies have been investigated, reported, and suggested by scientists to the livestock industry. Yet, the majority of studies have solely reported the impact of proposed strategies on the final product (methane production); the process of methane reduction has often been ambiguous. Thus, clarifying the process of methane reduction in the rumen is essential. Despite the positive effects on reducing enteric CH_4 emissions, the genetic selection method leads to decreased rumen volume and increased passage rate. Therefore, it may result in decreased ruminal fermentation and production of VFA, and finally will reduce the efficiency of microbial fermentation. Manipulation of the rumen microbiome seems to be a better method, but it requires further studies to investigate its different aspects.

Author Contributions: Conceptualization, V.P.; methodology, V.K.E., V.P. and S.E.; writing—original draft preparation, V.K.E., V.P. and S.E.; writing—review and editing, V.K.E., V.P. and S.E.; visualization, S.E. All authors have read and agreed to the published version of the manuscript.

Funding: This research received no external funding.

Institutional Review Board Statement: Not applicable.

Informed Consent Statement: Not applicable.

Data Availability Statement: Not applicable.

Conflicts of Interest: The authors declare no conflict of interest.

References

1. Kumari, S.; Fagodiya, R.K.; Hiloidhari, M.; Dahiya, R.P.; Kumar, A. Methane Production and Estimation from Livestock Husbandry: A Mechanistic Understanding and Emerging Mitigation Options. *Sci. Total Environ.* **2020**, *709*, 136135. [CrossRef] [PubMed]
2. Herrero, M.; Henderson, B.; Havlík, P.; Thornton, P.K.; Conant, R.T.; Smith, P.; Wirsenius, S.; Hristov, A.N.; Gerber, P.; Gill, M.; et al. Greenhouse Gas Mitigation Potentials in the Livestock Sector. *Nat. Clim. Chang.* **2016**, *6*, 452–461. [CrossRef]
3. De Congio, G.F.S.; Bannink, A.; Mayorga Mogollón, O.L.; Jaurena, G.; Gonda, H.; Gere, J.I.; Cerón-Cucchi, M.E.; Ortiz-Chura, A.; Tieri, M.P.; Hernández, O.; et al. Enteric Methane Mitigation Strategies for Ruminant Livestock Systems in the Latin America and Caribbean Region: A Meta-Analysis. *J. Clean. Prod.* **2021**, *312*, 127693. [CrossRef]
4. González-Recio, O.; López-Paredes, J.; Ouatahar, L.; Charfeddine, N.; Ugarte, E.; Alenda, R.; Jiménez-Montero, J.A. Mitigation of Greenhouse Gases in Dairy Cattle via Genetic Selection: 2. Incorporating Methane Emissions into the Breeding Goal. *J. Dairy Sci.* **2020**, *103*, 7210–7221. [CrossRef]
5. Pszczola, M.; Calus, M.P.L.; Strabel, T. Short Communication: Genetic Correlations between Methane and Milk Production, Conformation, and Functional Traits. *J. Dairy Sci.* **2019**, *102*, 5342–5346. [CrossRef]

6. Ornelas, L.T.C.; Silva, D.C.; Tomich, T.R.; Campos, M.M.; Machado, F.S.; Ferreira, A.L.; Maurício, R.M.; Pereira, L.G.R. Differences in Methane Production, Yield and Intensity and Its Effects on Metabolism of Dairy Heifers. *Sci. Total Environ.* **2019**, *689*, 1133–1140. [CrossRef]
7. Lan, W.; Yang, C. Ruminal Methane Production: Associated Microorganisms and the Potential of Applying Hydrogen-Utilizing Bacteria for Mitigation. *Sci. Total Environ.* **2019**, *654*, 1270–1283. [CrossRef] [PubMed]
8. Palangi, V.; Macit, M. Indictable Mitigation of Methane Emission Using Some Organic Acids as Additives Towards a Cleaner Ecosystem. *Waste Biomass Valor.* **2021**, *12*, 4825–4834. [CrossRef]
9. De Haas, Y.; Windig, J.J.; Calus, M.P.L.; Dijkstra, J.; de Haan, M.; Bannink, A.; Veerkamp, R.F. Genetic Parameters for Predicted Methane Production and Potential for Reducing Enteric Emissions through Genomic Selection. *J. Dairy Sci.* **2011**, *94*, 6122–6134. [CrossRef]
10. De Mulder, T.; Peiren, N.; Vandaele, L.; Ruttink, T.; De Campeneere, S.; van de Wiele, T.; Goossens, K. Impact of Breed on the Rumen Microbial Community Composition and Methane Emission of Holstein Friesian and Belgian Blue Heifers. *Livest. Sci.* **2018**, *207*, 38–44. [CrossRef]
11. Mercadante, M.E.Z.; de Melo Caliman, A.P.; Canesin, R.C.; Bonilha, S.F.M.; Berndt, A.; Frighetto, R.T.S.; Magnani, E.; Branco, R.H. Relationship between Residual Feed Intake and Enteric Methane Emission in Nellore Cattle. *Rev. Bras. Zootec.* **2015**, *44*, 255–262. [CrossRef]
12. Todd, R.W.; Altman, M.B.; Cole, N.A.; Waldrip, H.M. Methane Emissions from a Beef Cattle Feedyard during Winter and Summer on the Southern High Plains of Texas. *J. Environ. Qual.* **2014**, *43*, 1125–1130. [CrossRef]
13. Pickering, N.K.; Oddy, V.H.; Basarab, J.; Cammack, K.; Hayes, B.; Hegarty, R.S.; Lassen, J.; McEwan, J.C.; Miller, S.; Pinares-Patino, C.S.; et al. Animal Board Invited Review: Genetic Possibilities to Reduce Enteric Methane Emissions from Ruminants. *Animal* **2015**, *9*, 1431–1440. [CrossRef] [PubMed]
14. Engelke, S.W.; Daş, G.; Derno, M.; Tuchscherer, A.; Wimmers, K.; Rychlik, M.; Kienberger, H.; Berg, W.; Kuhla, B.; Metges, C.C. Methane Prediction Based on Individual or Groups of Milk Fatty Acids for Dairy Cows Fed Rations with or without Linseed. *J. Dairy Sci.* **2019**, *102*, 1788–1802. [CrossRef] [PubMed]
15. Leahy, S.C.; Kelly, W.J.; Ronimus, R.S.; Wedlock, N.; Altermann, E.; Attwood, G.T. Genome Sequencing of Rumen Bacteria and Archaea and Its Application to Methane Mitigation Strategies. *Animal* **2013**, *7*, 235–243. [CrossRef] [PubMed]
16. Colombini, S.; Rota Graziosi, A.; Parma, P.; Iriti, M.; Vitalini, S.; Sarnataro, C.; Spanghero, M. Evaluation of Dietary Addition of 2 Essential Oils from *Achillea moschata*, or Their Components (Bornyl Acetate, Camphor, and Eucalyptol) on in Vitro Ruminal Fermentation and Microbial Community Composition. *Anim. Nutr.* **2021**, *7*, 224–231. [CrossRef]
17. Niu, W.; He, Y.; Wang, H.; Xia, C.; Shi, H.; Cao, B.; Su, H. Effects of *Leymus chinensis* Replacement with Whole-Crop Wheat Hay on Blood Parameters, Fatty Acid Composition, and Microbiomes of Holstein Bulls. *J. Dairy Sci.* **2018**, *101*, 246–256. [CrossRef]
18. Feng, X.Y.; Dijkstra, J.; Bannink, A.; van Gastelen, S.; France, J.; Kebreab, E. Antimethanogenic Effects of Nitrate Supplementation in Cattle: A Meta-Analysis. *J. Dairy Sci.* **2020**, *103*, 11375–11385. [CrossRef]
19. Torres, R.N.S.; Paschoaloto, J.R.; Ezequiel, J.M.B.; da Silva, D.A.V.; Almeida, M.T.C. Meta-Analysis of the Effects of Essential Oil as an Alternative to Monensin in Diets for Beef Cattle. *Vet. J.* **2021**, *272*, 105659. [CrossRef]
20. McGinn, S.M.; Beauchemin, K.A.; Coates, T.; Colombatto, D. Methane Emissions from Beef Cattle: Effects of Monensin, Sunflower Oil, Enzymes, Yeast, and Fumaric Acid. *J. Anim. Sci.* **2004**, *82*, 3346–3356. [CrossRef]
21. Latham, E.A.; Pinchak, W.E.; Trachsel, J.; Allen, H.K.; Callaway, T.R.; Nisbet, D.J.; Anderson, R.C. Paenibacillus 79R4, a Potential Rumen Probiotic to Enhance Nitrite Detoxification and Methane Mitigation in Nitrate-Treated Ruminants. *Sci. Total Environ.* **2019**, *671*, 324–328. [CrossRef] [PubMed]
22. Torres, R.N.S.; Ghedini, C.P.; Paschoaloto, J.R.; da Silva, D.A.V.; Coelho, L.M.; Almeida Junior, G.A.; Ezequiel, J.M.B.; Machado Neto, O.R.; Almeida, M.T.C. Effects of Tannins Supplementation to Sheep Diets on Their Performance, Carcass Parameters and Meat Fatty Acid Profile: A Meta-Analysis Study. *Small Rumin. Res.* **2022**, *206*, 106585. [CrossRef]
23. Thompson, L.R.; Rowntree, J.E. INVITED REVIEW: Methane Sources, Quantification, and Mitigation in Grazing Beef Systems. *Appl. Anim. Sci.* **2020**, *36*, 556–573. [CrossRef]
24. Malmuthuge, N.; Griebel, P.J.; Guan, L.L. The Gut Microbiome and Its Potential Role in the Development and Function of Newborn Calf Gastrointestinal Tract. *Front. Vet. Sci.* **2015**, *2*, 36. [CrossRef] [PubMed]
25. Subharat, S.; Shu, D.; Zheng, T.; Buddle, B.M.; Janssen, P.H.; Luo, D.; Wedlock, D.N. Vaccination of Cattle with a Methanogen Protein Produces Specific Antibodies in the Saliva Which Are Stable in the Rumen. *Vet. Immunol. Immunopathol.* **2015**, *164*, 201–207. [CrossRef]
26. Wenner, B.A.; Wagner, B.K.; St-Pierre, N.R.; Yu, Z.T.; Firkins, J.L. Inhibition of Methanogenesis by Nitrate, with or without Defaunation, in Continuous Culture. *J. Dairy Sci.* **2020**, *103*, 7124–7140. [CrossRef]
27. Lassen, J.; Difford, G.F. Review: Genetic and Genomic Selection as a Methane Mitigation Strategy in Dairy Cattle. *Animal* **2020**, *14*, s473–s483. [CrossRef]
28. de Haas, Y.; Pszczola, M.; Soyeurt, H.; Wall, E.; Lassen, J. Invited Review: Phenotypes to Genetically Reduce Greenhouse Gas Emissions in Dairying. *J. Dairy Sci.* **2017**, *100*, 855–870. [CrossRef]

29. Grainger, C.; Clarke, T.; McGinn, S.M.; Auldist, M.J.; Beauchemin, K.A.; Hannah, M.C.; Waghorn, G.C.; Clark, H.; Eckard, R.J. Methane Emissions from Dairy Cows Measured Using the Sulfur Hexafluoride (SF6) Tracer and Chamber Techniques. *J. Dairy Sci.* **2007**, *90*, 2755–2766. [CrossRef]
30. Ricci, P.; Rooke, J.A.; Nevison, I.; Waterhouse, A. Methane Emissions from Beef and Dairy Cattle: Quantifying the Effect of Physiological Stage and Diet Characteristics. *J. Anim. Sci.* **2013**, *91*, 5379–5389. [CrossRef]
31. Rischewski, J.; Bielak, A.; Nürnberg, G.; Derno, M.; Kuhla, B. Rapid Communication: Ranking Dairy Cows for Methane Emissions Measured Using Respiration Chamber or GreenFeed Techniques during Early, Peak, and Late Lactation. *J. Anim. Sci.* **2017**, *95*, 3154–3159. [CrossRef] [PubMed]
32. Garnsworthy, P.C.; Craigon, J.; Hernandez-Medrano, J.H.; Saunders, N. On-Farm Methane Measurements during Milking Correlate with Total Methane Production by Individual Dairy Cows. *J. Dairy Sci.* **2012**, *95*, 3166–3180. [CrossRef] [PubMed]
33. Szalanski, M.; Kristensen, T.; Difford, G.; Lassen, J.; Buitenhuis, A.J.; Pszczola, M.; Løvendahl, P. Enteric Methane Emission from Jersey Cows during the Spring Transition from Indoor Feeding to Grazing. *J. Dairy Sci.* **2019**, *102*, 6319–6329. [CrossRef] [PubMed]
34. Lassen, J.; Løvendahl, P.; Madsen, J. Accuracy of Noninvasive Breath Methane Measurements Using Fourier Transform Infrared Methods on Individual Cows. *J. Dairy Sci.* **2012**, *95*, 890–898. [CrossRef] [PubMed]
35. Herd, R.M.; Arthur, P.F.; Donoghue, K.A.; Bird, S.H.; Bird-Gardiner, T.; Hegarty, R.S. Measures of Methane Production and Their Phenotypic Relationships with Dry Matter Intake, Growth, and Body Composition Traits in Beef Cattle. *J. Anim. Sci.* **2014**, *92*, 5267–5274. [CrossRef]
36. Jonker, A.; Hickey, S.M.; Rowe, S.J.; Janssen, P.H.; Shackell, G.H.; Elmes, S.; Bain, W.E.; Wing, J.; Greer, G.J.; Bryson, B.; et al. Genetic Parameters of Methane Emissions Determined Using Portable Accumulation Chambers in Lambs and Ewes Grazing Pasture and Genetic Correlations with Emissions Determined in Respiration Chambers1. *J. Anim. Sci.* **2018**, *96*, 3031–3042. [CrossRef]
37. Denninger, T.M.; Schwarm, A.; Dohme-Meier, F.; Münger, A.; Bapst, B.; Wegmann, S.; Grandl, F.; Vanlierde, A.; Sorg, D.; Ortmann, S.; et al. Accuracy of Methane Emissions Predicted from Milk Mid-Infrared Spectra and Measured by Laser Methane Detectors in Brown Swiss Dairy Cows. *J. Dairy Sci.* **2020**, *103*, 2024–2039. [CrossRef]
38. Tomkins, N.W.; McGinn, S.M.; Turner, D.A.; Charmley, E. Comparison of Open-Circuit Respiration Chambers with a Micrometeorological Method for Determining Methane Emissions from Beef Cattle Grazing a Tropical Pasture. *Anim. Feed Sci. Technol.* **2011**, *166–167*, 240–247. [CrossRef]
39. Denninger, T.M.; Dohme-Meier, F.; Eggerschwiler, L.; Vanlierde, A.; Grandl, F.; Gredler, B.; Kreuzer, M.; Schwarm, A.; Münger, A. Persistence of Differences between Dairy Cows Categorized as Low or High Methane Emitters, as Estimated from Milk Mid-Infrared Spectra and Measured by GreenFeed. *J. Dairy Sci.* **2019**, *102*, 11751–11765. [CrossRef]
40. Jonker, A.; Molano, G.; Antwi, C.; Waghorn, G.C. Enteric Methane and Carbon Dioxide Emissions Measured Using Respiration Chambers, the Sulfur Hexafluoride Tracer Technique, and a Greenfeed Head-Chamber System from Beef Heifers Fed Alfalfa Silage at Three Allowances and Four Feeding Frequencies. *J. Anim. Sci.* **2016**, *94*, 4326–4337. [CrossRef]
41. Manzanilla-Pech, C.I.V.; de Haas, Y.; Hayes, B.J.; Veerkamp, R.F.; Khansefid, M.; Donoghue, K.A.; Arthur, P.F.; Pryce, J.E. Genomewide Association Study of Methane Emissions in Angus Beef Cattle with Validation in Dairy Cattle. *J. Anim. Sci.* **2016**, *94*, 4151–4166. [CrossRef]
42. López-Paredes, J.; Goiri, I.; Atxaerandio, R.; García-Rodríguez, A.; Ugarte, E.; Jiménez-Montero, J.A.; Alenda, R.; González-Recio, O. Mitigation of Greenhouse Gases in Dairy Cattle via Genetic Selection: 1. Genetic Parameters of Direct Methane Using Noninvasive Methods and Proxies of Methane. *J. Dairy Sci.* **2020**, *103*, 7199–7209. [CrossRef]
43. Cabezas-Garcia, E.H.; Krizsan, S.J.; Shingfield, K.J.; Huhtanen, P. Between-Cow Variation in Digestion and Rumen Fermentation Variables Associated with Methane Production. *J. Dairy Sci.* **2017**, *100*, 4409–4424. [CrossRef]
44. Montes, F.; Meinen, R.; Dell, C.; Rotz, A.; Hristov, A.N.; Oh, J.; Waghorn, G.; Gerber, P.J.; Henderson, B.; Makkar, H.P.S.; et al. SPECIAL TOPICS-Mitigation of Methane and Nitrous Oxide Emissions from Animal Operations: II. A Review of Manure Management Mitigation Options. *J. Anim. Sci.* **2013**, *91*, 5070–5094. [CrossRef]
45. Richardson, C.M.; Nguyen, T.T.T.; Abdelsayed, M.; Moate, P.J.; Williams, S.R.O.; Chud, T.C.S.; Schenkel, F.S.; Goddard, M.E.; van den Berg, I.; Cocks, B.G.; et al. Genetic Parameters for Methane Emission Traits in Australian Dairy Cows. *J. Dairy Sci.* **2021**, *104*, 539–549. [CrossRef] [PubMed]
46. Difford, G.F.; Olijhoek, D.W.; Hellwing, A.L.F.; Lund, P.; Bjerring, M.A.; de Haas, Y.; Lassen, J.; Løvendahl, P. Ranking Cows' Methane Emissions under Commercial Conditions with Sniffers versus Respiration Chambers. *Acta Agric. Scand. A Anim. Sci.* **2018**, *68*, 25–32. [CrossRef]
47. Brask, M.; Weisbjerg, M.R.; Hellwing, A.L.F.; Bannink, A.; Lund, P. Methane Production and Diurnal Variation Measured in Dairy Cows and Predicted from Fermentation Pattern and Nutrient or Carbon Flow. *Animal* **2015**, *9*, 1795–1806. [CrossRef] [PubMed]
48. Robinson, D.L.; Goopy, J.P.; Hegarty, R.S.; Oddy, V.H.; Thompson, A.N.; Toovey, A.F.; Macleay, C.A.; Briegal, J.R.; Woodgate, R.T.; Donaldson, A.J.; et al. Genetic and Environmental Variation in Methane Emissions of Sheep at Pasture. *J. Anim. Sci.* **2014**, *92*, 4349–4363. [CrossRef] [PubMed]
49. Kittelmann, S.; Pinares-Patiño, C.S.; Seedorf, H.; Kirk, M.R.; Ganesh, S.; McEwan, J.C.; Janssen, P.H. Two Different Bacterial Community Types Are Linked with the Low-Methane Emission Trait in Sheep. *PLoS ONE* **2014**, *9*, e0103171. [CrossRef] [PubMed]

50. New Zealand Society of Animal Production Online Archive. Available online: https://www.nzsap.org/ (accessed on 28 September 2022).
51. Xiang, R.; McNally, J.; Rowe, S.; Jonker, A.; Pinares-Patino, C.S.; Oddy, V.H.; Vercoe, P.E.; McEwan, J.C.; Dalrymple, B.P. Gene Network Analysis Identifies Rumen Epithelial Cell Proliferation, Differentiation and Metabolic Pathways Perturbed by Diet and Correlated with Methane Production. *Sci. Rep.* **2016**, *6*, 39022. [CrossRef]
52. Xiang, R.; Dalrymple, B.P. Across-Experiment Transcriptomics of Sheep Rumen Identifies Expression of Lipid / Oxo-Acid Metabolism and Muscle Cell Junction Genes Associated With Variation in Methane-Related Phenotypes Differences Between the Two. *Front. Genet.* **2018**, *9*, 330. [CrossRef]
53. Swainson, N.; Muetzel, S.; Clark, H. Updated Predictions of Enteric Methane Emissions from Sheep Suitable for Use in the New Zealand National Greenhouse Gas Inventory. *Anim. Prod. Sci.* **2018**, *58*, 973–979. [CrossRef]
54. Renand, G.; Decruyenaere, V.; Maupetit, D.; Dozias, D. Methane and Carbon Dioxide Emission of Beef Heifers in Relation with Growth and Feed E Ffi Ciency. *Animals* **2019**, *3*, 1–17.
55. Andersen, T.O.; Kunath, B.J.; Hagen, L.H.; Arntzen, M.; Pope, P.B. Rumen Metaproteomics: Closer to Linking Rumen Microbial Function to Animal Productivity Traits. *Methods* **2021**, *186*, 42–51. [CrossRef] [PubMed]
56. Ribeiro, D.M.; Salama, A.A.K.; Vitor, A.C.M.; Argüello, A.; Moncau, C.T.; Santos, E.M.; Caja, G.; de Oliveira, J.S.; Balieiro, J.C.C.; Hernández-Castellano, L.E.; et al. The Application of Omics in Ruminant Production: A Review in the Tropical and Sub-Tropical Animal Production Context. *J. Proteom.* **2020**, *227*, 103905. [CrossRef]
57. Guzman, C.E.; Bereza-Malcolm, L.T.; De Groef, B.; Franks, A.E. Presence of Selected Methanogens, Fibrolytic Bacteria, and Proteobacteria in the Gastrointestinal Tract of Neonatal Dairy Calves from Birth to 72 Hours. *PLoS ONE* **2015**, *10*, e0133048. [CrossRef]
58. Yáñez-Ruiz, D.R.; Abecia, L.; Newbold, C.J. Manipulating Rumen Microbiome and Fermentation through Interventions during Early Life: A Review. *Front. Microbiol.* **2015**, *6*, 1133. [CrossRef]
59. Rey, M.; Enjalbert, F.; Monteils, V. Establishment of Ruminal Enzyme Activities and Fermentation Capacity in Dairy Calves from Birth through Weaning. *J. Dairy Sci.* **2012**, *95*, 1500–1512. [CrossRef]
60. Pitta, D.; Indugu, N.; Narayan, K.; Hennessy, M. Symposium Review: Understanding the Role of the Rumen Microbiome in Enteric Methane Mitigation and Productivity in Dairy Cows. *J. Dairy Sci.* **2022**, *105*, 8569–8585. [CrossRef]
61. Friedman, N.; Jami, E.; Mizrahi, I. Compositional and Functional Dynamics of the Bovine Rumen Methanogenic Community across Different Developmental Stages. *Environ. Microbiol.* **2017**, *19*, 3365–3373. [CrossRef]
62. Arshad, M.A.; ul Hassan, F.; Rehman, M.S.; Huws, S.A.; Cheng, Y.; Din, A.U. Gut Microbiome Colonization and Development in Neonatal Ruminants: Strategies, Prospects, and Opportunities. *Anim. Nutr.* **2021**, *7*, 883–895. [CrossRef] [PubMed]
63. Palma-Hidalgo, J.M.; Jiménez, E.; Popova, M.; Morgavi, D.P.; Martín-García, A.I.; Yáñez-Ruiz, D.R.; Belanche, A. Inoculation with Rumen Fluid in Early Life Accelerates the Rumen Microbial Development and Favours the Weaning Process in Goats. *Anim. Microbiome* **2021**, *3*, 11. [CrossRef]
64. Weimer, P.J. Redundancy, Resilience, and Host Specificity of the Ruminal Microbiota: Implications for Engineering Improved Ruminal Fermentations. *Front. Microbiol.* **2015**, *6*, 296. [CrossRef] [PubMed]
65. Henderson, G.; Cox, F.; Ganesh, S.; Jonker, A.; Young, W.; Janssen, P.H.; Abecia, L.; Angarita, E.; Aravena, P.; Arenas, G.N.; et al. Rumen Microbial Community Composition Varies with Diet and Host, but a Core Microbiome Is Found across a Wide Geographical Range. *Sci. Rep.* **2015**, *5*, 14567. [CrossRef]
66. Paillard, D.; McKain, N.; Chaudhary, L.C.; Walker, N.D.; Pizette, F.; Koppova, I.; McEwan, N.R.; Kopečný, J.; Vercoe, P.E.; Louis, P.; et al. Relation between Phylogenetic Position, Lipid Metabolism and Butyrate Production by Different *Butyrivibrio*-like Bacteria from the Rumen. *Antonie Van Leeuwenhoek Int. J. Gen. Mol. Microbiol.* **2007**, *91*, 417–422. [CrossRef] [PubMed]
67. Beauchemin, K.A.; Ungerfeld, E.; Gruninger, R. *Control of Methanogenesis in Dairy Animals*, 3rd ed.; Elsevier: Amsterdam, The Netherlands, 2021; Volume 2, ISBN 9780128187661.
68. Sharifi, M.; Taghizadeh, A.; Hosseinkhani, A.; Palangi, V.; Maclt, M.; Salem, A.Z.M.; Elghndour, M.M.M.Y.; Abachi, S. Influence of Nitrate Supplementation on In-Vitro Methane Emission, Milk Production, Ruminal Fermentation, and Microbial Methanotrophs in Dairy Cows Fed at Two Forage Levels. *Anim. Sci.* **2022**, *22*, 1–22. [CrossRef]
69. Kliem, K.E.; Humphries, D.J.; Kirton, P.; Givens, D.I.; Reynolds, C.K. Differential Effects of Oilseed Supplements on Methane Production and Milk Fatty Acid Concentrations in Dairy Cows. *Animal* **2019**, *13*, 309–317. [CrossRef] [PubMed]
70. Wischer, G.; Boguhn, J.; Steingaß, H.; Schollenberger, M.; Hartung, K.; Rodehutscord, M. Effect of Monensin on in Vitro Fermentation of Silages and Microbial Protein Synthesis. *Arch. Anim. Nutr.* **2013**, *67*, 219–234. [CrossRef] [PubMed]
71. Valdez-Vazquez, I.; Poggi-Varaldo, H.M. Hydrogen Production by Fermentative Consortia. *Renew. Sustain. Energy Rev.* **2009**, *13*, 1000–1013. [CrossRef]
72. Schären, M.; Drong, C.; Kiri, K.; Riede, S.; Gardener, M.; Meyer, U.; Hummel, J.; Urich, T.; Breves, G.; Dänicke, S. Differential Effects of Monensin and a Blend of Essential Oils on Rumen Microbiota Composition of Transition Dairy Cows. *J. Dairy Sci.* **2017**, *100*, 2765–2783. [CrossRef] [PubMed]
73. Mammi, L.M.E.; Guadagnini, M.; Mechor, G.; Cainzos, J.M.; Fusaro, I.; Palmonari, A.; Formigoni, A. The Use of Monensin for Ketosis Prevention in Dairy Cows during the Transition Period: A Systematic Review. *Animals* **2021**, *11*, 1988. [CrossRef]

74. Ipharraguerre, I.R.; Clark, J.H. Usefulness of Ionophores for Lactating Dairy Cows: A Review. *Anim. Feed Sci. Technol.* **2003**, *106*, 39–57. [CrossRef]
75. Saborío-Montero, A.; Gutiérrez-Rivas, M.; Goiri, I.; Atxaerandio, R.; García-Rodriguez, A.; López-Paredes, J.; Jiménez-Montero, J.A.; González-Recio, O. Rumen Eukaryotes Are the Main Phenotypic Risk Factors for Larger Methane Emissions in Dairy Cattle. *Livest. Sci.* **2022**, *263*, 105023. [CrossRef]
76. Wang, Y.; Nan, X.; Zhao, Y.; Jiang, L.; Wang, H.; Hua, D.; Zhang, F.; Wang, Y.; Liu, J.; Yao, J.; et al. Dietary Supplementation with Inulin Improves Lactation Performance and Serum Lipids by Regulating the Rumen Microbiome and Metabolome in Dairy Cows. *Anim. Nutr.* **2021**, *7*, 1189–1204. [CrossRef] [PubMed]
77. Zhao, X.H.; Gong, J.M.; Zhou, S.; Liu, C.J.; Qu, M.R. The Effect of Starch, Inulin, and Degradable Protein on Ruminal Fermentation and Microbial Growth in Rumen Simulation Technique. *Ital. J. Anim. Sci.* **2014**, *13*, 189–195. [CrossRef]
78. Tian, K.; Liu, J.; Sun, Y.; Wu, Y.; Chen, J.; Zhang, R.; He, T.; Dong, G. Effects of Dietary Supplementation of Inulin on Rumen Fermentation and Bacterial Microbiota, Inflammatory Response and Growth Performance in Finishing Beef Steers Fed High or Low-Concentrate Diet. *Anim. Feed Sci. Technol.* **2019**, *258*, 114299. [CrossRef]
79. Maia, M.R.G.; Chaudhary, L.C.; Figueres, L.; Wallace, R.J. Metabolism of Polyunsaturated Fatty Acids and Their Toxicity to the Microflora of the Rumen. *Antonie Van Leeuwenhoek Int. J. Gen. Mol. Microbiol.* **2007**, *91*, 303–314. [CrossRef] [PubMed]
80. Maia, M.R.; Chaudhary, L.C.; Bestwick, C.S.; Richardson, A.J.; McKain, N.; Larson, T.R.; Graham, I.A.; Wallace, R.J. Toxicity of Unsaturated Fatty Acids to the Biohydrogenating Ruminal Bacterium, *Butyrivibrio fibrisolvens*. *BMC Microbiol.* **2010**, *10*, 52. [CrossRef] [PubMed]
81. Burdick, M.; Zhou, M.; Guan, L.L.; Oba, M. Effects of Medium-Chain Fatty Acid Supplementation on Performance and Rumen Fermentation of Lactating Holstein Dairy Cows. *Animal* **2022**, *16*, 100491. [CrossRef] [PubMed]
82. Danielsson, R.; Dicksved, J.; Sun, L.; Gonda, H.; Müller, B.; Schnürer, A.; Bertilsson, J. Methane Production in Dairy Cows Correlates with Rumen Methanogenic and Bacterial Community Structure. *Front. Microbiol.* **2017**, *8*, 226. [CrossRef]
83. Pandey, D.; Helene, H.; Dhakal, R.; Aryal, N.; Prakash, S.; Sapkota, R.; Olaf, M.; Novoa-garrido, M.; Khanal, P. Interspecies and Seasonal Variations in Macroalgae from the Nordic Region: Chemical Composition and Impacts on Rumen Fermentation and Microbiome Assembly. *J. Clean. Prod.* **2022**, *363*, 132456. [CrossRef]
84. Wallace, R.J.; Rooke, J.A.; McKain, N.; Duthie, C.A.; Hyslop, J.J.; Ross, D.W.; Waterhouse, A.; Watson, M.; Roehe, R. The Rumen Microbial Metagenome Associated with High Methane Production in Cattle. *BMC Genom.* **2015**, *16*, 839. [CrossRef] [PubMed]
85. Hernandez-Sanabria, E.; Goonewardene, L.A.; Wang, Z.; Durunna, O.N.; Moore, S.S.; Guan, L.L. Impact of Feed Efficiency and Diet on Adaptive Variations in the Bacterial Community in the Rumen Fluid of Cattle. *Appl. Environ. Microbiol.* **2012**, *78*, 1203–1214. [CrossRef] [PubMed]
86. Xue, M.Y.; Sun, H.Z.; Wu, X.H.; Guan, L.L.; Liu, J.X. Assessment of Rumen Bacteria in Dairy Cows with Varied Milk Protein Yield. *J. Dairy Sci.* **2019**, *102*, 5031–5041. [CrossRef] [PubMed]
87. Vargas-Bello-Pérez, E.; Cancino-Padilla, N.; Romero, J.; Garnsworthy, P.C. Quantitative Analysis of Ruminal Bacterial Populations Involved in Lipid Metabolism in Dairy Cows Fed Different Vegetable Oils. *Animal* **2016**, *10*, 1821–1828. [CrossRef]
88. Gruninger, R.J.; Zhang, X.M.; Smith, M.L.; Kung, L.; Vyas, D.; McGinn, S.M.; Kindermann, M.; Wang, M.; Tan, Z.L.; Beauchemin, K.A. Application of 3-Nitrooxypropanol and Canola Oil to Mitigate Enteric Methane Emissions of Beef Cattle Results in Distinctly Different Effects on the Rumen Microbial Community. *Anim. Microbiome* **2022**, *4*, 35. [CrossRef]
89. Yang, S.L.; Bu, D.P.; Wang, J.Q.; Hu, Z.Y.; Li, D.; Wei, H.Y.; Zhou, L.Y.; Loor, J.J. Soybean Oil and Linseed Oil Supplementation Affect Profiles of Ruminal Microorganisms in Dairy Cows. *Animal* **2009**, *3*, 1562–1569. [CrossRef]
90. Zhou, R.; Wu, J.; Lang, X.; Liu, L.; Casper, D.P.; Wang, C.; Zhang, L.; Wei, S. Effects of Oregano Essential Oil on in Vitro Ruminal Fermentation, Methane Production, and Ruminal Microbial Community. *J. Dairy Sci.* **2020**, *103*, 2303–2314. [CrossRef]
91. Tapio, I.; Snelling, T.J.; Strozzi, F.; Wallace, R.J. The Ruminal Microbiome Associated with Methane Emissions from Ruminant Livestock. *J. Anim. Sci. Biotechnol.* **2017**, *8*, 7. [CrossRef]
92. Vasta, V.; Daghio, M.; Cappucci, A.; Buccioni, A.; Serra, A.; Viti, C.; Mele, M. Invited Review: Plant Polyphenols and Rumen Microbiota Responsible for Fatty Acid Biohydrogenation, Fiber Digestion, and Methane Emission: Experimental Evidence and Methodological Approaches. *J. Dairy Sci.* **2019**, *102*, 3781–3804. [CrossRef]
93. Pitta, D.W.; Pinchak, W.E.; Dowd, S.; Dorton, K.; Yoon, I.; Min, B.R.; Fulford, J.D.; Wickersham, T.A.; Malinowski, D.P. Longitudinal Shifts in Bacterial Diversity and Fermentation Pattern in the Rumen of Steers Grazing Wheat Pasture. *Anaerobe* **2014**, *30*, 11–17. [CrossRef] [PubMed]
94. Mu, Y.Y.; Qi, W.P.; Zhang, T.; Zhang, J.Y.; Mao, S.Y. Gene Function Adjustment for Carbohydrate Metabolism and Enrichment of Rumen Microbiota with Antibiotic Resistance Genes during Subacute Rumen Acidosis Induced by a High-Grain Diet in Lactating Dairy Cows. *J. Dairy Sci.* **2021**, *104*, 2087–2105. [CrossRef] [PubMed]
95. Rico, J.L.; Reardon, K.F.; de Long, S.K. Inoculum Microbiome Composition Impacts Fatty Acid Product Profile from Cellulosic Feedstock. *Bioresour. Technol.* **2021**, *323*, 124532. [CrossRef]
96. Suen, G.; Weimer, P.J.; Stevenson, D.M.; Aylward, F.O.; Boyum, J.; Deneke, J.; Drinkwater, C.; Ivanova, N.N.; Mikhailova, N.; Chertkov, O.; et al. The Complete Genome Sequence of Fibrobacter Succinogenes S85 Reveals a Cellulolytic and Metabolic Specialist. *PLoS ONE* **2011**, *6*, e0018814. [CrossRef]

97. Castells, L.; Bach, A.; Aris, A.; Terré, M. Effects of Forage Provision to Young Calves on Rumen Fermentation and Development of the Gastrointestinal Tract. *J. Dairy Sci.* **2013**, *96*, 5226–5236. [CrossRef]
98. Kim, Y.H.; Nagata, R.; Ohtani, N.; Ichijo, T.; Ikuta, K.; Sato, S. Effects of Dietary Forage and Calf Starter Diet on Ruminal PH and Bacteria in Holstein Calves during Weaning Transition. *Front. Microbiol.* **2016**, *7*, 1575. [CrossRef] [PubMed]
99. Lin, X.; Wang, J.; Hou, Q.; Wang, Y.; Hu, Z.; Shi, K.; Yan, Z.; Wang, Z. Effect of Hay Supplementation Timing on Rumen Microbiota in Suckling Calves. *Microbiologyopen* **2018**, *7*, 430. [CrossRef] [PubMed]
100. Jami, E.; Israel, A.; Kotser, A.; Mizrahi, I. Exploring the Bovine Rumen Bacterial Community from Birth to Adulthood. *ISME J.* **2013**, *7*, 1069–1079. [CrossRef] [PubMed]

methane

MDPI

Article

Evaluation of Rumen Methane Emission in Sahiwal and Gir Calves Supplemented with Combination of Methanogenic Inhibitors

Rachala Dinesh Reddy [1,†], Parul Chaudhary [2,†], Nitin Tyagi [1], Madhu Mohini [1] and Goutam Mondal [1,*]

1 Animal Nutrition Division, National Dairy Research Institute, Karnal 132001, Haryana, India
2 School of Agriculture, Graphic Era Hill University, Dehradun 248002, Uttarakhand, India
* Correspondence: gmondal1075@gmail.com or goutam.mondal@icar.gov.in
† These authors contributed equally to this work and share first authorship.

Abstract: Methane is one of the main greenhouse gases emitted by ruminants around the world. It is essential to investigate novel approaches to increasing animal production while reducing greenhouse gas emissions from ruminants. This study was conducted to examine the effect of methane inhibitors, such as nitrate, linseed oil, and anthraquinone, on nutritional digestibility, rumen fermentation processes, and methane emission in Sahiwal and Gir cattle calves. Twelve calves (6–12 months old), six of each Sahiwal and Gir breed, were selected and divided into four groups; Sahiwal control (C) and treated (T) calves; Gir control (C) and treated calves (T) of three calves each based on average body weight. Switch over a design was used as for periods 1 and 2. Animals in all groups were fed chopped oat fodder, wheat straw, and a concentrate mixture. Additionally, treated groups were fed a ration with potassium nitrate (1%), linseed oil (0.5%), and anthraquinone (4 ppm). The results revealed that the addition of methane inhibitors had no impact on nutrient intake and apparent digestibility. The levels of propionate, ammonia nitrogen, and total nitrogen were increased significantly ($p < 0.05$), while butyrate decreased in the treated groups of both breeds. However, there was no change in acetate and pH between the groups. Methane emission (g/d) was lower ($p < 0.05$) in the treated groups as compared to the control group. This study concludes that supplementation of methane inhibitors in calves feed can be utilized to lower methane emissions without affecting the intake and digestibility of nutrients. Combining diverse dietary mitigation strategies could be an effective way to mitigate methane emissions to reduce global warming while minimizing any negative impacts on ruminants to accomplish sustainable animal production.

Keywords: Gir calves; Sahiwal; methane mitigation; nutrient utilization

1. Introduction

Livestock is responsible for more than 15% of anthropogenic methane emissions via enteric fermentation. Rumen fermentation in ruminants produces methane, which results in a 2–12% loss of food energy while also contributing to greenhouse gas (GHG) burdens [1]. Ruminal methane is produced by methanogenic archaea by combining CO_2 and hydrogen (H+). Hydrogen must be removed because its buildup hinders a variety of biological processes needed to keep the rumen environment healthy. Methanogenesis appears to be the main pathway for H+ elimination in the rumen, despite the presence of alternative pathways, such as acetogenesis, propionogenesis, sulphate, and nitrate reduction [2]. Ruminant methane emissions also represent precious metabolic energy that is lost to the animals and must be replaced by their diet, which is another issue related to these emissions that goes beyond global warming. Thus, enteric methane reduction is being considered a promising technique for improving ruminant feed efficiency and lowering GHG emissions.

Many natural substances have been investigated to determine their efficacy in lowering methane emission under in vitro and in vivo conditions while optimizing rumen fermentation [3,4]. Utilizing substances that consume hydrogen is one of the most important ways to reduce methanogenesis. Nitrate is a common hydrogen-consuming chemical used to minimize methane production, and a non-protein nitrogen supply for rumen microbes [5]. The application of nitrate along with cysteamine had a positive impact on rumen fermentation while lowering the total gas and methane production by increasing the fiber degrading and hydrogen using microbes [6]. Long-term feeding of coated nitrate has also been shown in vivo to continually lower the intestinal methane emissions of grazing steers [7]. Adding sodium nitrate (0.11–0.44 kg body weight) did not negatively affect the water buffaloes milk yield, fatty acid, and rumen methanogen/*Butyrivibrio* population associated with biohydrogenation; however, it did increase the rumen volatile fatty acid and microbial diversity [8]. Application of nitrate reduced methane production linearly while having moderate impacts on rumen fermentation and have no effect on nutrient digestion in dairy cows [9].

It is also well known that supplemental lipids in the diet have been shown to change the rumen environment, affecting microbial fermentation and *volatile fatty acid* (VFA) synthesis. Linseed is one of the most intriguing lipid supplements in Malaysia and China because of its lower content of linoleic and saturated fatty acids and contains higher levels of alfa-linoleic acid and omega 3-fatty acid [10,11]. These substances have beneficial bioactivity for animal health due to their anti-inflammatory, antioxidative, and lipid-modulating effects [12]. Linseed has been used in ruminant nutrition to increase milk production and its quality [13]. It was observed that supplementing with linseed oil is a viable way to improve the composition of fatty acid in goat milk without affecting the animal performance parameters [14]. A naturally occurring aromatic chemical molecule, such as anthraquinone (9, 10-dioxoanthracene, AQ) is present in several plants. There are various natural sources of AQ, such as *Cassia fistula*, *Aloe succtrina*, *Senna alexandri*, and *Artemisia scoparia*, etc. AQ compounds are employed as laxatives, but they have antibacterial, antiviral, and antiparasitic effects. The rhubarb compounds, such as phthalic acid, isobutyl octadecyl ester, and di iso-octyl phthalate, were shown to have more specificity towards the binding site of methyl-coenzyme M reductase and may be an effective methanogen inhibitor [15]. However, studies on the combined impact of nitrate along with linseed oil and anthraquinone on the alteration of ruminal fermentation parameters are scarce. However, it is difficult to use a single component to impart the National environment program to herders, so a combination may be used. Consequently, the current study was conducted to assess the impact of methanogen inhibitors (nitrate, linseed oil and anthraquinone) on nutritional digestibility, rumen fermentation processes, and methane emission in Sahiwal and Gir calves.

2. Results

2.1. Chemical and Mineral Composition of Feed

The chemical composition (% DM) of feed provided to experimental animals is presented in Table 1. The oat green, wheat straw and concentrate mixture had 16.61, 91.91, and 89.26% of DM, respectively. Oats, wheat straw, and concentrate mixture CP contents were 8.76, 3.57, and 19.32%, respectively. The EE content was 2.63, 1.36, and 3.82%, respectively.

2.2. Body Weight and Intake of Nutrients

The body weight and nutrient intake data during the experiment are depicted in Table 2. Sahiwal (C and T) and Gir (C and T) had no change in body weight (159.16, 161.08, 167.82, and 163.05, respectively). It was observed that the addition of methane inhibitors had no impact on nutrient intake. The dry matter intake (DMI, kg/d) was similar among breeds and was 4.14, 3.97, 4.48, and 4.18 kg/d in Sahiwal (C), Sahiwal (T), Gir (C), and Gir (T), respectively. The CP and TDN intake were 0.53, 1.58 in Sahiwal (C), 0.54 and 1.45 in Sahiwal (T), 0.55, 1.64 in Gir (C), and 0.52 and 1.54 in Gir (T), respectively. In addition, the daily intake of EE, NDF, and ADF was 0.13, 1.99 and 1.36 kg/d in Sahiwal (C), 0.12, 1.87, and 1.28 in Sahiwal (T), 0.14, 2.21, and 1.51 in Gir (C) and 0.12, 1.97, and 1.33 in Gir (T), respectively.

Table 1. Chemical composition of feedstuffs during the experiment.

Parameter	Oats	Wheat Straw	Concentrate Mixture
Dry matter (%)	16.61	90.91	89.26
Organic matter (%)	90.41	88.75	89.55
Crude protein (%)	8.76	3.57	19.32
Ether extract (%)	2.63	1.36	3.82
Neutral detergent fiber (%)	59.36	77.92	23.45
Acid detergent fiber (%)	40.05	55.33	14.17
Crude fiber (%)	27.04	40.65	3.73
Total ash (%)	9.58	11.25	10.44

Table 2. Growth performance, nutrient intake, and digestibility in different groups: Sahiwal control (C), Sahiwal treated (T), Gir control (C), and Gir treated calves (T).

Parameter	Sahiwal (C)	Sahiwal (T)	Gir (C)	Gir (T)	*p*-Value
Body weight (kg)	159.16 ± 14.99	161.08 ± 3.55	167.82 ± 19.09	163.05 ± 10.03	0.971
Dry Matter intake through oats (kg/d)	1.30 ± 0.21	1.27 ± 0.08	1.45 ± 0.20	1.38 ± 0.08	0.864
Dry matter intake through wheat straw (kg/d)	0.88 ± 0.16	0.72 ± 0.09	1.07 ± 0.20	0.85 ± 0.14	0.527
Dry matter intake through Concentrate mix. (kg/d)	1.97 ± 0.03	1.98 ± 0.02	1.96 ± 0.02	1.95 ± 0.02	0.809
Total dry matter intake (kg/d)	4.14 ± 0.38	3.97 ± 0.14	4.48 ± 0.43	4.18 ± 0.24	0.720
Dry matter intake (kg/100 kg BW)	2.61 ± 0.04	2.46 ± 0.05	2.70 ± 0.07	2.57 ± 0.06	0.072
Dry matter intake (g/kgW $^{0.75}$)	92.24 ± 2.04	87.62 ± 2.15	96.30 ± 2.16	91.69 ± 2.17	0.770
Organic matter intake (kg)	3.71 ± 0.35	3.55 ± 0.13	4.01 ± 0.39	3.74 ± 0.21	0.732
Organic matter intake (kg/100 kg BW)	2.34 ± 0.03	2.20 ± 0.05	2.41 ± 0.06	2.30 ± 0.06	0.071
Crude protein intake (kg/d)	0.53 ± 0.02	0.52 ± 0.01	0.55 ± 0.02	0.52 ± 0.02	0.506
Crude protein intake (kg/100 kg BW)	0.35 ± 0.02	0.32 ± 0.00	0.34 ± 0.03	0.32 ± 0.01	0.723
Ether extract intake (kg)	0.13 ± 0.01	0.12 ± 0.01	0.14 ± 0.01	0.12 ± 0.01	0.238
Ether extract intake (kg/100 kg BW)	0.08 ± 0.00	0.07 ± 0.01	0.09 ± 0.00	0.07 ± 0.01	0.076
Neutral detergent fiber intake (kg)	1.99 ± 0.31	1.87 ± 0.12	2.21 ± 0.35	1.97 ± 0.13	0.804
Neutral detergent fiber (kg/100 kg BW)	1.22 ± 0.08	1.16 ± 0.06	1.28 ± 0.08	1.21 ± 0.04	0.615
Acid detergent fiber intake (kg)	1.36 ± 0.17	1.28 ± 0.06	1.51 ± 0.20	1.33 ± 0.12	0.724
Acid detergent fiber (kg/100 kg BW)	0.84 ± 0.03	0.79 ± 0.03	0.89 ± 0.03	0.81 ± 0.04	0.195
Crude fiber intake (kg)	0.76 ± 0.19	0.73 ± 0.09	0.87 ± 0.20	0.83 ± 0.06	0.904
Crude fiber intake (kg/100 kg BW)	0.44 ± 0.08	0.45 ± 0.05	0.48 ± 0.08	0.52 ± 0.03	0.833
Total digestible nutrients intake (kg/d)	2.52 ± 0.25	2.33 ± 0.15	2.75 ± 0.30	2.50 ± 0.18	0.651
Total digestible nutrients intake (kg/100 kg BW)	1.58 ± 0.03	1.45 ± 0.07	1.64 ± 0.03	1.54 ± 0.05	0.117

Values in tables were the means ± standard.

2.3. *Apparent Digestibility Coefficients of Nutrients*

The digestibility coefficients of DM, OM, CP, EE, NDF, and ADF were 63.19, 64.31, 63.46, 74.60, 55.46, and 42.98% in the control calves, 61.40, 62.60, 63.37, 73.25, 54.43, and 41.47 in treated Sahiwal calves, 62.28, 64.80, 63.32, 75.30, 57.06, and 42.97 in control and 62.99, 63.77, 62.84, 73.69, 54.66, and 40.90 in treated Gir calves, respectively (Table 3). It was observed that there was no discernible change in the apparent digestibility of the Sahiwal and Gir breeds in either control or treated groups.

Table 3. Digestibility coefficients (%) of various nutrients in the different groups of calves: Sahiwal control (C), Sahiwal treated (T), Gir control (C), and Gir treated calves (T).

Parameter	Sahiwal (C)	Sahiwal (T)	Gir (C)	Gir (T)	*p*-Value
Dry matter (%)	63.19 ± 1.29	61.40 ± 1.75	64.28 ± 1.20	62.99 ± 2.15	0.689
Organic matter (%)	64.31 ± 1.26	62.60 ± 1.71	64.80 ± 1.26	63.77 ± 2.13	0.812
Crude protein (%)	63.46 ± 1.23	63.37 ± 1.14	63.32 ± 0.92	62.84 ± 1.94	0.989
Ether extract (%)	74.60 ± 0.81	73.25 ± 1.04	75.30 ± 1.13	73.69 ± 1.51	0.613
Neutral detergent fiber (%)	55.46 ± 1.85	54.43 ± 2.08	57.06 ± 2.06	54.66 ± 3.26	0.867
Acid detergent fiber (%)	42.98 ± 2.14	41.47 ± 2.57	42.97 ± 2.59	40.90 ± 3.55	0.934

Values in tables were the means ± standard.

2.4. Rumen Fermentation Parameters in Different Groups

The levels of acetate, propionate, and butyrate were 63.91, 19.24, and 12.30% in control; 62.90, 21.86, and 11.23% in treated Sahiwal calves; 62.79, 18.75, and 12.71% in control and 61.69, 21.34, and 11.68% in Gir calves treated with methane inhibitors, respectively (Table 4). Acetate values did not vary significantly, although there was an increase in propionate and a decrease in butyrate between the treated groups of Sahiwal and Gir as compared to the control. The ratio of acetate: propionate ($p > 0.05$) was 3.33, 2.89, 3.35, and 2.89 in Sahiwal and Gir calves. In both breeds, treatment with methane inhibitors caused a considerable decline in the A:P ratio. The NH3N (mg/dL) and total N (g/dL) values considerably increased in treated groups. The values for NH3N and N were 18.10 and 72.33 in the control group of Sahiwal calves, 19.88 and 74.66 in treated Sahiwal (T), 19.69 and 109.66 in Gir control calves, and 21.65 and 112.00 in treated Gir calves, respectively.

Table 4. Rumen fermentation patterns in the different groups of calves: Sahiwal control (C), Sahiwal treated (T), Gir control (C), and Gir treated calves (T).

Parameter	Sahiwal (C)	Sahiwal (T)	Gir (C)	Gir (T)	*p*-Value
Acetate (%)	63.91 ± 0.86	62.90 ± 0.60	62.79 ± 0.65	61.69 ± 0.21	0.130
Propionate (%)	19.24 ^{a} ± 0.38	21.86 ^{b} ± 0.47	18.75 ^{a} ± 0.37	21.34 ^{b} ± 0.18	<0.011
Butyrate (%)	12.30 ^{ab} ± 0.29	11.23 ^{a} ± 0.40	12.71 ^{b} ± 0.37	11.68 ^{ab} ± 0.29	0.033
A:P	3.33 ^{b} ± 0.11	2.89 ^{a} ± 0.08	3.35 ^{b} ± 0.07	2.89 ^{a} ± 0.01	<0.013
NH_3N (mg/dL)	18.10 ^{a} ± 0.21	19.88 ^{ab} ± 0.36	19.69 ^{ab} ± 0.48	21.65 ^{b} ± 0.94	0.034
Total N (g/dL)	72.33 ^{a} ± 3.90	74.66 ^{a} ± 1.47	109.66 ^{b} ± 5.32	112.00 ^{b} ± 5.11	<0.014
pH	6.58 ± 0.02	6.57 ± 0.03	6.61 ± 0.03	6.63 ± 0.04	0.485

Means bearing different superscripts ^{a} and ^{b} in the same row differ significantly ($p < 0.05$).

2.5. Enteric Methane Emission in Different Groups

Methane emission (g/d) was lower ($p < 0.01$) in the treated groups (51.90 and 61.38 in Sahiwal and Gir, respectively) as compared to the control group (65.39 and 74.55 in Sahiwal and Gir, respectively). Methane emissions per kg dry matter intake (DMI) and digestible dry matter intake (DDMI) ranged between 13–17 g and 21–27 g in the four groups. The organic matter intake (OMI) and digestible dry organic matter intake (DOMI) levels were between 14–19 and 23–30 g, respectively. Methane emissions (g/kg CPI) were significantly ($p < 0.05$) greater in control groups of both breeds Sahiwal C (123.63) and Gir C (135.92) than the treated groups Sahiwal T (100.64) and Gir T (118.20), respectively (Table 5). Due to the addition of methane inhibitors, the CH4 (g/d) was lower in the treated groups as compared to the control groups. The CH4 energy loss as GEI, DEI, and MEI was less than 20% and 14% in Sahiwal T and Gir T compared to the control groups. Gir calves lost more energy as methane as compared to Sahiwal, indicating that Sahiwal calves had a greater response toward the methane inhibitor supplementation.

Table 5. Rumen methane emission in different groups: Sahiwal control (C), Sahiwal treated (T), Gir control (C), and Gir treated calves (T).

Parameter	Sahiwal (C)	Sahiwal (T)	Gir (C)	Gir (T)	*p*-Value
CH_4 (g/d)	65.39 ^{b} ± 2.70	51.90 ^{a} ± 2.88 (−20.63%)	74.55 ^{c} ± 1.01 (+12.28%)	61.38 ^{b} ± 1.43 (−17.66%)	<0.010
CH4 (g/kg DMI)	16.55 ± 1.72	13.18 ± 0.86 (−20.36%)	17.37 ± 1.54 (+4.72%)	14.96 ± 1.06 (−13.87%)	0.162
CH4 (g/kg DDMI)	26.24 ± 2.79	21.58 ± 1.56 (−17.76%)	27.22 ± 2.84 (+3.60%)	23.90 ± 1.80 (−12.20%)	0.340
CH4 (g/kg OMI)	18.50 ± 1.95	14.72 ± 0.96 (−20.43%)	19.44 ± 1.78 (+4.84%)	16.70 ± 1.14 (−14.09%)	0.163
CH_4 (g/kg DOMI)	28.83 ± 3.12	23.61 ± 1.68 (−18.10%)	30.27 ± 3.28 (+4.76%)	26.33 ± 1.92 (−13.01%)	0.313
CH_4 (g/kg CPI)	123.63 ^{ab} ± 7.24	100.64 ^{a} ± 6.02 (−18.59%)	135.92 ^{b} ± 3.19 (+9.04%)	118.20 ^{ab} ± 6.45 (−13.03%)	0.040
CH_4 (g/kg TDNI)	27.39 ± 2.96	22.59 ± 1.59 (−17.52%)	28.78 ± 3.06 (+4.83%)	25.14 ± 1.77 (−12.65%)	0.324
CH_4 energy loss as %					
Gross energy intake (MJ/d)	4.95 ± 0.47	3.93 ± 0.23	5.20 ± 0.41	4.46 ± 0.28	0.140
Digestible energy intake (MJ/d)	8.08 ^{ab} ± 0.65	6.44 ^{a} ± 0.36	8.63 ^{b} ± 0.53	7.39 ^{ab} ± 0.41	0.057
Metabolizable energy intake (MJ/d)	9.73 ± 0.79	7.76 ± 0.44	10.38 ± 0.65	8.91 ± 0.50	0.062

Means bearing different superscripts $^{a, b}$ and ^{c} in the same row differ significantly ($p < 0.05$).

3. Discussion

3.1. Feed Intake and Digestibility

Sahiwal and Gir are two milch cattle breeds popular in Northern and Western India due to their distinct appearance, ability to withstand high temperatures, and resistance to parasite infestation. This part of the country is characterized by extreme weather conditions, and more than 50% of the human population is dependent on animal protein/other nutrients from milk and milk products. There has been no research conducted to compare the nutrient intake, digestibility, and methane emission in Gir and Sahiwal breed cattle under similar feeding conditions. GHG emission is a global issue, and livestock has a huge impact on a country's GHG inventory making up 60% of the agricultural sector. Therefore, it is crucial to monitor methane emission and rumen fermentation when feeding these popular breeds in India. As India has a diverse population of livestock, enteric methane emissions may not be similar even under similar feeding regimens. Again, combinations of lipids, plant secondary metabolites, and chemical additives have an effect on the mitigation of methane from ruminants. Lipid feeding, such as linseed oil, reduced methane emission by reducing the quantity of OM fermented in the rumen and by exerting direct toxic effects of FA on rumen methanogens [16]. Nitrate may reduce methane emission by competing with methanogenesis for accessible hydrogen in the rumen [9]. AQ appears to affect the methanogen methyl-coenzyme M by inhibiting electron transport during methane production. Thus, the combined impact of nitrate, linseed oil, and anthraquinone on nutrient digestibility and methane emission has been investigated in this study.

Voluntary feed intake is a crucial criterion that has a significant impact on animal productivity and welfare; hence, it is important to check for any response on DMI when evaluating a feed supplement and additives. Feed intake was unaltered in the current experiment, indicating that the diet palatability was not impacted by the methane inhibitors in the control and treated groups of both Sahiwal and Gir breeds. This suggests that the addition of methane inhibitors had no impact on palatability and nutrient intake. Studies have shown inconsistent results regarding the impact of linseed oil and nitrate supplementation on intake and nutritional digestibility. Supplementation of flaxseed oil enhanced the OM and CP digestibility without altering the feed intake. The oil used was roughly 2.4% of the feed intake, which is below the permissible limit for ruminal microbiota activity, which is not high to depress consumption [17]. Kholif et al. [18] also observed that adding flaxseed oil to the diets of goats (20 mL/d) boosted feed utilization and milk production. When linseed oil (2, 3, and 4%) was added to lactating cows' diet, there was no difference in the amount of DM or apparent digestibility of nutrients [19]. It was also observed that nitrate and potassium (2.6 and 4%) in the sheep diet had no impact on the consumption of DM [20,21]. Feeding nitrate has two advantages as it reduces methane emissions while also providing ruminants with non-protein nitrogen. The inclusion of anthraquinone and chloroform during the rearing had no effect on the DMI and growth of dairy calves [22]. In this study, the addition of methane inhibitors had no effect on the apparent digestibility. Supplementation of linseed oil and rubber seed oil could improve nutrient digestibility and rumen fermentation by increasing the vaccenic acid, cis-9 trans-11 conjugated linoleic, and α-linolenic acid composition in dairy lactating cows [23]. The variation could be attributed to species, level of production, and basal diet of the experimental animals.

3.2. Rumen Fermentation Parameters

Rumen fermentation produces excess hydrogen, which must be eliminated from the rumen in order for the fermentation process and microbial development to continue efficiently [24]. Methanogenesis is a crucial mechanism in the rumen for disposing of electrons generated during fermentation, and suppression of methanogens may result in an accumulation of hydrogen that could restrict rumen fermentation. Thus, it is preferable to shift the reducing equivalents from carbon dioxide and hydrogen to processes other than methanogenesis, such as acetate, and to selectively stimulate fermentation to increase propionate synthesis [25].

There was a significant increase in propionate, ammonia nitrogen, and total nitrogen, and a decrease in butyrate was observed in treated groups of both breeds. This may be due to the shift of H^+ concentrations in the rumen towards propionate and the direct inhibition of methanogens which stimulate acetate production. An increase in H_+ concentrations in the rumen could change feed fermentation pathways to produce a lesser amount of acetate and more propionate, shifting the methanogen community composition away from foremost *Methanobrevibacter* spp. [26]. AQ at 4 ppm increased propionate concentration, which could be a reason for using part of the hydrogen that would be used for microbial lipid production [27]. However, there was no change in acetate and pH was observed between the groups due to the additive effects of methanogenic inhibitors on methanogenesis, or they may be affecting the acetogenic bacterial population [28]. Feed containing plant oils may make the rumen more acidic, lowering the ruminal pH. Ruminal pH and acetate percentage were both decreased ($p < 0.05$) by flaxseed oil in lactating Nubian goats [17]. The impact of nitrate on ruminal fermentation is consistent with the increased rumen pH seen in treated groups after 6 hr of feeding in beef steers [29]. An elevation in ruminal pH might be because the experiments used different types of feed in different studies. Nitrate supplementation enhanced the population of *Prevotella* spp., which may have contributed to the higher propionate proportion [30]. Nitrate supplementation in dairy cows increased dissolved H concentration, microbial N, and propionate molar percentage while reducing ammonia concentrations in rumen fluid and methane emissions [31]. Encapsulated nitrate (ENS) is a feed supplement that reduces *Methanobrevibacter* abundance in the rumen by persistently affecting intestinal methane emission in grazing steers. Furthermore, ENS can stimulate fumarate-reducing and lactate-producing bacteria, lowering acetate synthesis during rumen fermentation [7]. Linseed oil and nitrate addition to ruminant diets also change the VFA profile so that there is more propionate and less acetate [32,33].

3.3. Enteric Methane Emission

In this study, it was observed that supplementation of methane inhibitors to the diet significantly decreased methane emission in the treated groups as compared to control in both breeds approx 20%. Methane reduction may be due to the nitrate and sulfate reduction which is thermodynamically preferred over CO_2 reduction, giving nitrate and sulfate reducing bacteria a competitive edge over methanogens as hydrogen sinks. It was found that feeding Holstein steers with dietary nitrate with six levels (0–3% of feed DM) resulted linear decrease in methane emission [34]. However, supplementation of nitrate and fumarate with a plant leaves-based diet in sheep had no impact on intake/nutrient utilization, blood profile, microbial N, and rumen fermentation parameters [35]. Klop et al. [36] found that the addition of nitrate (1% DM) decreased methane production (g/kg of DMI) by 7.6–10% in dairy lactating cows. It was observed that the addition of linseed oil (4%) with red clover silage-based diet decreased intestinal methane (about 9%), energy losses (about 11%), and N excretion (8%), respectively [37]. Supplementation of linseed oil to ruminant diets had a positive impact on fermentation processes and decreased methane production while increasing propionic acid without interfering with feed digestion [38]. Linseed diet reduced the ruminal acetate proportion, archaea to bacteria ratio, and reduced methane emission in lactating cows [39]. However, under field conditions, a combination of inhibitors may be more useful than an individual due to the synergistic effect of the additive. One product may act on the methanogenic population, the other on its enzyme system, or third act as an H+ sink in the system.

4. Material and Methods

4.1. Experimental Location

The experiment was carried out at the National Dairy Research Institute Livestock Research Center (LRC), Karnal, Haryana, India, which is located at an altitude of 29°42′ N and 76°54′ E. The minimum and maximum ambient temperature range from near freezing in the winter (4 °C) to 46 °C in the summer, with a diurnal variation of 15–20 °C. The research practice and maintenance were in accordance with the Institute Animal Ethics Committee (IAEC) standards, and consent was obtained from the committee with IAEC approval no. 43-IAEC-18-16.

4.2. Experimental Design

Twelve calves (6–12 months old), six of each breed Sahiwal and Gir, were chosen randomly from the herd at LRC, Karnal, and divided into four groups; Sahiwal control (C) and treated (T) calves; Gir control (C) and treated calves (T) of three calves each based on average body weight. The switch over design was used for periods 1 and 2 (Figure 1). All of the animals were fed in accordance with the ICAR 2013 feeding standard [40]. Animals in all groups were fed chopped oat fodder, wheat straw, and a concentrate mixture. In the control and treated groups, the level of concentrate, oats, and wheat straw were 30:40:30. Additionally, the treated groups were given a meal that included potassium nitrate (1%), linseed oil (0.5%), and anthraquinone (4 ppm). For period 2, as previously stated, the animals were transferred from control to treatment and vice versa.

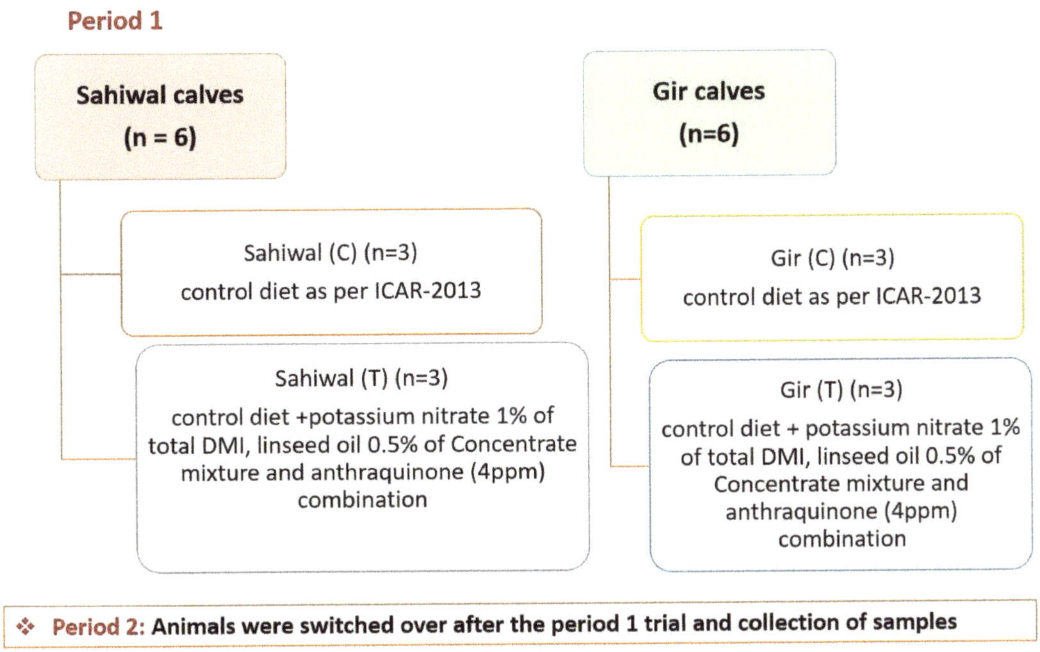

Figure 1. Experimental design for different groups.

4.3. Digestibility Trial

Digestibility trial of seven days collection period was undertaken once in each trial during the last month of the experimental period to assess the nutrient digestibility. The experiment involved housing each animal separately in a shed while making provisions for the quantitative collection of feed/fodders, residues left, and feces. Before beginning the collection, animals were allowed two days to acclimatize in the shed.

4.4. Sampling, Processing, and Storage of Feed Samples

The samples of different feeds offered (concentrate mixture, green fodder, and wheat straw) residues, if any, were taken daily for dry matter (DM) estimation during the metabolism trial. These samples were pooled at the end of the collection period and ground to pass through a 1 mm sieve and stored in airtight containers. The samples were analyzed for proximate principles (OM, CP, Ash, and EE) and cell wall constituents (NDF and ADF). Feces voided during 24 h were collected daily for seven days and weighed at 9:00 am daily. After thorough mixing, 1/100 of the total sample on a weight basis was kept for DM estimation. Dried pooled fecal samples were ground to pass through a 1 mm sieve size and analyzed for proximate principles and cell wall constituents as per standard procedures. For N estimation of feces, an aliquot of 1/500 of total voided feces were collected daily for seven days and stored in plastic containers containing 25 mL of H_2SO_4 solution.

4.5. Analysis of Feed, Faeces, and Urine Samples

Dried samples of feed offered, residues, and feces were examined for chemical composition, such as dry and organic matter (DM, OM), ethyl ether (EE), crude protein (CP) [41], and cell wall fractions, such as neutral and acid detergent fiber (NDF, ADF) [42].

4.6. Rumen Fermentation Parameters

Rumen liquor was drawn through a stomach tube from various parts of the animal's rumen just before feeding, thoroughly mixed, and strained. The pH of the strained rumen liquor (SRL) was then promptly recorded. Rumen liquor was placed at −20 °C for NH_3-N and total nitrogen estimation. Samples were chilled while being acidified with a few drops of H_2SO_4 (25%) for total and individual volatile fatty acid estimation.

4.6.1. Individual Fatty Acid Estimation (IVFA)

Rumen liquor frozen at 4 °C for 24 h and centrifuged at 3000 rpm. The collected supernatant (4 mL) was treated with meta-phosphoric acid (1 mL) and stored at 4 °C overnight, and centrifuged at 3000 rpm for 10 min to estimate the IVFA via gas chromatography (Nucon 5700, India) fitted with a flame ionization detector (FID) and a stainless-steel column with Chromosorb 101. The injection port, column, and detector were all adjusted at 200, 180, and 210 °C, respectively. The carrier gas (N_2) flow rate through the column was 40 mL/min, whereas the flow rate of H_2 and air via FID were 20 and 300 mL/min, respectively. The sample (3 μL) was injected through the injection port by means of a Hamilton syringe. Diverse VFAs were identified based on the area covered by their retention times on the monitor, and their concentration was determined by comparing the peak areas of the standard [43].

4.6.2. Ammonia and Total Nitrogen Estimation

Rumen liquor (5 mL) and NaOH (5 mL, 40%) were taken in the distillation assembly for the estimation of ammonia. The distillate was collected and titrated against H_2SO_4 (0.01 N) in a conical flask containing boric acid solution (2%, 10 mL). Rumen liquor (2 mL) was placed in a Kjeldahl digestion flask and digested by adding concentrated H_2SO_4 (10 mL). Digested material was diluted with distilled water to a volume of 100 mL. An aliquot of the digested sample (10 mL) was placed in the distillation apparatus, along with NaOH (40%, 15 mL). The distillate was collected in a conical flask with a boric acid solution (2%, 20 mL), and a mixed indicator was taken in the flask and titrated against H_2SO_4 (0.01 N) for the total N estimation [3,4].

4.7. Estimation of Methane Production Using SF6 Method

The enteric methane (CH_4) production was estimated using the SF_6 tracer gas technique for five days. In this method, the permeation tubes were prepared by filling them with a specific amount of SF_6 and inserted into the rumen with a known release rate. Each animal was fitted with a halter and capillary tube attached to an evacuated sampling canister set to fill halfway in 24 h. Sample from the animal's mouth and nose was taken as the vacuum in the sampling canister gradually dissipated. Background CH_4 and SF_6

concentrations were measured for each day by placing one sampling kit in a naturally ventilated house. The amount of CH_4 and SF_6 in animal samples were then corrected for background concentration. Following sample collection, the canister was pressurized with nitrogen and the concentration of SF_6 was determined using gas chromatography (Nucon 5700), which was equipped with an electron capture detector (250 °C) and a 3.3 m molecular sieve column. To estimate CH_4 concentration, another gas chromatograph was equipped with a flame ionization detector (100 °C) and a stainless-steel column packed with Porapak-Q. In both gas chromatographs, the injector and column were set at 50 and 40 °C, respectively [44].

4.8. Statistical Analysis

One-way analysis of variance was used to analyze the data for nutrient digestibility and methane emission with the help of SPSS software version 9.3. A general linear model was used to differentiate the treatment effect, and the level of significance was considered at 5%. The model used for analysis is $Y_{ij} = \mu + t_i + e_{ij}$, where y_{ij} is the observation on *i*th treatment for *j*th unit, μ is the grand mean, t_i is the effect of *i*th treatment, e_{ij} is random error distributed normally, and *i* is dependent on zero and constant variance. No random effects were considered as all the animals were picked from the herd based on their age and body weight.

5. Conclusions

This study concludes that the addition of methane inhibitors (nitrate, linseed oil, and anthraquinone) to dairy feed can be utilized to lower methane emissions without affecting the digestibility in both Sahiwal and Gir breeds. Combining diverse dietary mitigation strategies could be an effective way to reduce methane emissions while minimizing any negative impacts on calves health and performance to achieve sustainable animal production. However, to attain the optimum results, a thorough study of the rumen microbial population with different concentrations of methane inhibitors and AO's potential health risks is required to scale up at the national level.

Author Contributions: R.D.R. performed all experiments. P.C. wrote the manuscript and helped in data analysis. N.T. and M.M. helped in editing the manuscript. G.M. contributed to the experimental concept, design, and final editing of the manuscript. All authors have read and agreed to the published version of the manuscript.

Funding: Ministry of Environment, Forest and Climate Change, GoI funded the research work under 'Updatation of GHG inventory from Indian livestock'.

Institutional Review Board Statement: Approval was obtained from the Institute Animal Ethics committee with IAEC number 43-IAEC-18-16. This statement have been included in ethics declarations.

Informed Consent Statement: Not applicable.

Data Availability Statement: The study materials related to this study are included in this article.

Acknowledgments: We thank the Director, National Dairy Research Institute, Karnal, Haryana, India, for providing facilities to conduct this research.

Conflicts of Interest: The authors declare no conflict of interest.

References

1. Appuhamy, J.; France, J.; Kebreab, E. Models for predicting enteric methane emissions from dairy cows in North America, Europe, and Australia and New Zealand. *Glob. Chang. Biol.* **2016**, *22*, 3039–3056. [CrossRef]
2. Choudhury, P.K.; Jena, R.; Tomar, S.K.; Puniya, A.K. Reducing Enteric Methanogenesis through Alternate Hydrogen Sinks in the Rumen. *Methane* **2022**, *1*, 320–341. [CrossRef]
3. Leitanthem, V.K.; Chaudhary, P.; Bhakat, M.; Mohini, M.; Mondal, G. Impact of *Moringa oleifera* on rumen fermentation and methane emission under in vitro condition. *AMB Expr.* **2022**, *12*, 141. [CrossRef] [PubMed]
4. Leitanthem, V.K.; Chaudhary, P.; Maiti, S.; Mohini, M.; Mondal, G. Impact of *Moringa oleifera* Leaves on Nutrient Utilization, Enteric Methane Emissions, and Performance of Goat Kids. *Animals* **2023**, *13*, 97. [CrossRef] [PubMed]

5. Zu, H.; Xu, J.; Cong, Y. Reducing rumen methane emission through regulating rumen microorganisms by adding hydrogen-consuming compounds. *Chin. J. Anim. Sci.* **2019**, *31*, 4967–4972.
6. Guo, Y.; Hassan, F.-U.; Li, M.; Xie, H.; Peng, L.; Tang, Z.; Yang, C. Effect of Sodium Nitrate and Cysteamine on *In Vitro* Ruminal Fermentation, Amino Acid Metabolism and Microbiota in Buffalo. *Microorganisms* **2022**, *10*, 2038. [CrossRef]
7. Granja-Salcedo, Y.T.; Fernandes, R.M.; Araujo, R.C.D.; Kishi, L.T.; Berchielli, T.T.; Resende, F.D.D.; Berndt, A.; Siqueir, G.R. Long-term encapsulated nitrate supplementation modulates rumen microbial diversity and rumen fermentation to reduce methane emission in grazing steers. *Front. Microbiol.* **2019**, *10*, 614. [CrossRef]
8. Xie, F.; Tang, Z.; Liang, X.; Wen, C.; Li, M.; Guo, Y.; Peng, K.; Yang, C. Sodium nitrate has no detrimental effect on milk fatty acid profile and rumen bacterial population in water buffaloes. *AMB Expr.* **2022**, *12*, 11. [CrossRef]
9. Olijhoek, D.W.; Hellwing, A.L.F.; Brask, M.; Weisbjerg, M.R.; Højberg, O.; Larsen, M.K.; Dijkstra, J.; Erlandsen, E.J.; Lund, P. Effect of dietary nitrate level on enteric methane production, hydrogen emission, rumen fermentation, and nutrient digestibility in dairy cows. *J. Dairy Sci.* **2016**, *99*, 6191–6205. [CrossRef]
10. Abuelfatah, K.; Zuki, A.B.; Goh, Y.; Sazili, A.Q.; Abubakr, A. Effects of feeding whole linseed on ruminal fatty acid composition and microbial population in goats. *Anim. Nutr.* **2016**, *2*, 323–328. [CrossRef] [PubMed]
11. Xu, L.; Wei, Z.; Guo, B.; Bai, R.; Liu, J.; Li, Y.; Sun, W.; Jiang, X.; Li, X.; Pi, Y. Flaxseed Meal and Its Application in Animal Husbandry: A Review. *Agriculture* **2022**, *12*, 2027. [CrossRef]
12. Parikh, M.; Maddaford, T.G.; Austria, J.A.; Aliani, M.; Netticadan, T.; Pierce, G.N. Dietary Flaxseed as a Strategy for Improving Human Health. *Nutrients* **2019**, *11*, 1171. [CrossRef] [PubMed]
13. Musco, N.; Tudisco, R.; Esposito, G.; Iommelli, P.; Totakul, P.; D'Aniello, B.; Lombardi, P.; Amato, R.; Wanapat, M.; Infascelli, F. Effects of Linseed Supplementation on Milk Production, Composition, Odd- and Branched-Chain Fatty Acids, and on Serum Biochemistry in Cilentana Grazing Goats. *Animals* **2022**, *12*, 783. [CrossRef] [PubMed]
14. Gómez-Cortés, P.; Cívico, A.; de la Fuente, M.A.; Núñez Sánchez, N.; Peña Blanco, F.; Martínez Marín, A.L. Effects of dietary concentrate composition and linseed oil supplementation on the milk fatty acid profile of goats. *Animal* **2018**, *12*, 2310–2317. [CrossRef] [PubMed]
15. Arokiyaraj, S.; Stalin, A.; Shin, H. Anti-methanogenic effect of rhubarb (*Rheum* spp.)—An in silico docking studies on methyl-coenzyme M reductase (MCR). *Saudi J. Biol. Sci.* **2019**, *26*, 1458–1462. [CrossRef]
16. Kholif, S.M.; Morsy, T.A.; Matloup, O.H.; Ebeid, H.M.; Kholif, A.M. Effects of crushed linseed or linseed oil supplementation on performance of dairy goats and fatty acid profile in milk. *Life Sci. J.* **2015**, *12*, 94–99. [CrossRef]
17. Kholif, A.E.; Morsy, T.A.; Abdo, M.M. Crushed flaxseed versus flaxseed oil in the diets of Nubian goats: Effect on feed intake, digestion, ruminal fermentation, blood chemistry, milk production, milk composition and milk fatty acid profile. *Anim. Feed. Sci. Technol.* **2018**, *244*, 66–75. [CrossRef]
18. Kholif, A.E.; Morsy, T.A.; Abd El Tawab, A.M.; Anele, U.Y.; Galyean, M.L. Effect of supplementing diets of Anglo-Nubian goats with soybean and flaxseed oils on lactational performance. *J. Agric. Food Chem.* **2016**, *64*, 6163–6170. [CrossRef]
19. Benchaar, C.; Romero-Pérez, G.A.; Chouinard, P.Y.; Hassanat, F.; Eugene, M.; Petit, H.V.; Côrtes, C. Supplementation of increasing amounts of linseed oil to dairy cows fed total mixed rations: Effects on digestion, ruminal fermentation characteristics, protozoal populations, and milk fatty acid composition. *J. Dairy Sci.* **2012**, *95*, 4578–4590. [CrossRef]
20. Van Zijderveld, S.M.; Gerrits, W.J.J.; Apajalahti, J.A.; Newbold, J.R.; Dijkstra, J.; Leng, R.A.; Perdok, H.B. Nitrate and sulfate: Effective alternative hydrogen sinks for mitigation of ruminal methane production in sheep. *J. Dairy Sci.* **2010**, *93*, 5856–5866. [CrossRef]
21. Nolan, J.V.; Hegarty, R.S.; Hegarty, J.; Godwin, I.R.; Woodgate, R. Effects of dietary nitrate on rumen fermentation, methane production and digesta kinetics in sheep. *Anim. Prod. Sci.* **2010**, *50*, 801–806. [CrossRef]
22. Cristobal-Carballo, O.; McCoard, S.A.; Cookson, A.L.; Ganesh, S.; Lowe, K.; Laven, R.A.; Muetzel, S. Effect of Methane Inhibitors on Ruminal Microbiota During Early Life and Its Relationship With Ruminal Metabolism and Growth in Calves. *Front. Microbiol.* **2021**, *12*, 710914. [CrossRef] [PubMed]
23. Pi, Y.; Ma, L.; Pierce, K.; Wang, H.; Xu, J.; Bu, D. Rubber seed oil and flaxseed oil supplementation alter digestion, ruminal fermentation and rumen fatty acid profile of dairy cows. *Animal* **2019**, *13*, 2811–2820. [CrossRef] [PubMed]
24. Kelly, W.J.; Mackie, R.I.; Attwood, G.T.; Janssen, P.H.; McAllister, T.A.; Leahy, S.C. Hydrogen and formate production and utilisation in the rumen and the human colon. *Anim. Microbiome* **2022**, *4*, 22. [CrossRef]
25. Ungerfeld, E.M. Metabolic Hydrogen Flows in Rumen Fermentation: Principles and Possibilities of Interventions. *Front. Microbiol.* **2020**, *11*, 589. [CrossRef] [PubMed]
26. Martinez-Fernandez, G.; Denman, S.E.; Yang, C.; Cheung, J.; Mitsumori, M.; Mcsweeney, C.S. Methane inhibition alters the microbial community, hydrogen flow, and fermentation response in the rumen of cattle. *Front. Microbiol.* **2016**, *7*, 1122. [CrossRef]
27. Ebrahimi, S.H.; Mohini, M.; Singhal, K.K.; Miri, V.H.; Tyagi, A.K. Evaluation of complementary effects of 9,10-anthraquinone and fumaric acid on methanogenesis and ruminal fermentation in vitro. *Arch. Anim. Nutr.* **2011**, *65*, 267–277. [CrossRef]
28. Garcia-Lopez, P.M.; Kung, L.; Odom, J.J.M. In vitro inhibition of microbial methane production by 9,10-anthraquinone. *J. Anim. Sci.* **1996**, *74*, 2276–2284. [CrossRef]
29. Lee, C.; Araujo, R.C.; Koenig, K.M.; Beauchemin, K.A. Effects of encapsulated nitrate on growth performance, carcass characteristics, nitrate residues in tissues, and enteric methane emissions in beef steers: Finishing phase. *J. Anim. Sci.* **2017**, *95*, 3712–3726.

30. Wang, M.; Wang, R.; Xie, T.; Janssen, P.H.; Sun, X.; Beauchemin, K.A.; Tan, Z.; Tan, G.M. Shifts in rumen fermentation and microbiota are associated with dissolved ruminal hydrogen concentrations in lactating dairy cows fed different types of carbohydrates. *J. Nutr.* **2016**, *146*, 1714–1721. [CrossRef]
31. Wang, R.; Wang, M.; Ungerfeld, E.M.; Zhang, X.M.; Long, D.L.; Mao, H.X.; Deng, J.P.; Bannink, A.; Tan, Z.L. Nitrate improves ammonia incorporation into rumen microbial protein in lactating dairy cows fed a low-protein diet. *J. Dairy Sci.* **2018**, *101*, 9789–9799. [CrossRef] [PubMed]
32. Ivan, M.; Petit, H.V.; Chiquette, J.; Wright, A.-D.G. Rumen fermentation and microbial population in lactating dairy cows receiving diets containing oilseeds rich in C-18 fatty acids. *Br. J. Nutr.* **2013**, *109*, 1211–1218. [CrossRef] [PubMed]
33. Guyader, J.; Eugène, M.; Meunier, B.; Doreau, M.; Morgavi, D.P.; Silberberg, M.; Rochette, Y.; Gerard, C.; Loncke, C.; Martin, C. Additive methane-mitigating effect between linseed oil and nitrate fed to cattle. *J. Anim. Sci.* **2015**, *93*, 3564–3577. [CrossRef]
34. Newbold, J.R.; van Zijderveld, S.M.; Hulshof, R.B.A.; Fokkink, W.B.; Leng, R.A.; Terencio, P.; Powers, J.; Van Adrichem, P.S.J.; Paton, N.D.; Perdok, H.B. The effect of incremental levels of dietary nitrate on methane emissions in Holstein steers and performance in Nelore bulls. *J. Anim. Sci.* **2014**, *92*, 5032–5040. [CrossRef]
35. Pal, K.; Patra, A.K.; Sahoo, A.; Soren, N.M. Effects of nitrate and fumarate in tree leaves-based diets on nutrient utilization, rumen fermentation, microbial protein supply and blood profiles in sheep. *Livest. Sci.* **2015**, *172*, 5–15. [CrossRef]
36. Klop, G.; Hatew, B.; Bannink, A.; Dijkstra, J. Feeding nitrate and docosahexaenoic acid affects enteric methane production and milk fatty acid composition in lactating dairy cows. *J. Dairy Sci.* **2016**, *99*, 1161–1172. [CrossRef]
37. Benchaar, C.; Hassanat, F.; Martineau, R.; Gervais, R. Linseed oil supplementation to dairy cows fed diets based on red clover silage or corn silage: Effects on methane production, rumen fermentation, nutrient digestibility, N balance, and milk production. *J. Dairy Sci.* **2015**, *98*, 7993–8008. [CrossRef]
38. Vargas, J.E.; Andrés, S.; López-Ferreras, L.; Snelling, T.J.; Yáñez-Ruíz, D.R.; García-Estrada, C.; López, S. Dietary supplemental plant oils reduce methanogenesis from anaerobic microbial fermentation in the rumen. *Sci. Rep.* **2020**, *10*, 1613. [CrossRef] [PubMed]
39. van Gastelen, S.; Visker, M.H.P.W.; Edwards, J.E.; Antunes-Fernandes, E.C.; Hettinga, K.A.; Alferink, S.J.J.; Hendriks, W.H.; Bovenhuis, H.; Smidt, H.; Dijkstra, J. Linseed oil and DGAT1 K232A polymorphism: Effects on methane emission, energy and nitrogen metabolism, lactation performance, ruminal fermentation, and rumen microbial composition of Holstein-Friesian cows. *J. Dairy Sci.* **2017**, *100*, 8939–8957. [CrossRef]
40. ICAR. Nutrient requirements of cattle and buffalo. In *Nutrient Requirements of Animals*; Indian Council of Agricultural Research: New Delhi, Indian, 2013.
41. AOAC. Association of Official Analytical Chemist. In *Official Methods of Analysis*, 21st ed.; Association of Official Analytical Chemists: Washington, DC, USA, 2005; Volume 222.
42. Van Soest, P.J.; Robertson, J.B.; Lewis, B.A. Methods for dietary fiber, neutral detergent fiber, and nonstarch polysaccharides in relation to animal nutrition. *J. Dairy Sci.* **1991**, *74*, 3583–3597. [CrossRef]
43. Erwin, E.S.; Marco, G.J.; Emery, E.M. Volatile fatty acid analyses of blood and rumen fluid by gas chromatography. *J. Dairy Sci.* **1961**, *44*, 1768–1771. [CrossRef]
44. Johnson, K.; Huyler, M.; Westberg, H.; Lamb, B.; Zimmerman, P. Measurement of methane emissions from ruminant livestock using a sulfur hexafluoride tracer technique. *Environ. Sci. Tech.* **1994**, *28*, 359–362. [CrossRef] [PubMed]

methane

MDPI

Article

Modulating Natural Methane Release from Rumen Fermentation through the Use of *Ficus glomerata* Leaf Tannins in Murrah Buffalo (*Bubalus bubalis*)

Ram Kumar Singh, Avijit Dey * and Mala Singh

Division of Animal Nutrition and Feed Technology, ICAR-Central Institute for Research on Buffaloes, Hisar 125 001, India

* Correspondence: avijit.dey@icar.gov.in or avijitcirb@gmail.com; Tel.: +91-9728-334810; Fax: +91-1662-275004

Abstract: Enteric fermentation is one of the largest contributors of methane release to the environment from the livestock sector. Plant bioactive compounds can modulate rumen fermentation for reduced methanogenesis and fatty acid biohydrogenation. The present study investigates the effects of tannin extract from *Ficus glomerata* (FG) leaves on the rumen fermentation, methanogenesis, feed digestibility and fatty acid biohydrogenation of a total mixed ration with the aim of developing a feed supplement for enhanced livestock production and product quality with lower methane emission. The tannin extract (70% aqueous acetone extract) of FG leaves in the total mixed ration (oat hay/concentrate mixture; 1:1) was studied at four graded dose regimens (0.0 (control), 0.25 mL (FG-0.25), 0.50 mL (FG-0.50) and 1.0 mL (FG-1.0) per 60 mL of buffered rumen fluid) in three replicates for each treatment in a radio-frequency-based automatic gas production system (ANKOM-RF) at 39 °C for 24 h following the standard in vitro gas production protocol. The total gas production (mL or mL/g incubated dry matter (DM)) was gradually reduced ($p < 0.01$) at dose levels of FG-0.50 and FG-1.0; however, it remained intermediary and comparable ($p > 0.05$) for FG-0.25 with the control and FG-0.50. Compared to the control, the methane concentration (%) in the head space gas, as well as the total methane production (mL or mL/g DM incubated, or mL/g DM digested), were found to be gradually reduced ($p < 0.01$) with increasing doses (0.25–1.0 mL) of FG extract. The reduced ($p < 0.05$) feed degradability at higher levels (0.50–1.0 mL) of FG extract supplementation and the comparative ($p > 0.05$) effects with the control at a lower level of supplementation (FG-0.25) are suggestive of the dose-responsive detrimental effects of tannins on fibrolytic microbes in the rumen. However, the ammonia concentration decreased ($p < 0.05$) in all of the incubations compared to the control. Among the volatile fatty acids, acetate remained comparable ($p > 0.05$) with enhanced ($p < 0.05$) propionate at a lower dose (FG-0.25); however, a dose-dependent reduction was evident at higher dose levels (FG-0.50 and FG-1.0). The production of stearic acid (C18:0), which is a product of the rumen biohydrogenation process, was reduced ($p < 0.05$), irrespective of the concentration of the FG extract. Compared to the control, the concentration of *t*-vaccenic acid (C18:1), which is a precursor of conjugated linoleic acid (CLA) in animal products, was increased in all the FG-extract-supplemented groups. It may be concluded that *Ficus glomerata* leaf tannins can modulate rumen fermentation for reduced methanogenesis and fatty acid biohydrogenation in a total mixed ration. As a higher level of inclusion negatively affects feed digestibility, a lower dose (0.25 mL FG extract per 60 mL fermentation fluid or 4.17 mL FG extract per L of fermentation fluid) is suggested to achieve desirable effects on methane abatement (30%) and an improvement in fatty acid profiles in animal products.

Keywords: enteric methane emission; rumen biohydrogenation; *Ficus glomerata*; leaf tannins; buffalo

1. Introduction

The rumen is a natural fermenter in livestock and is responsible for degrading inedible feedstuffs by means of a diverse population of microorganisms to produce volatile fatty acids, which are a source of energy for the purposes of animal maintenance and production. The byproducts of this fermentation process are carbon dioxide and hydrogen, which form methane using the archaea that are present in the rumen and are emitted in the environment, mainly via eructation [1]. Globally, ruminants are one of the major sources of anthropogenic methane and nitrous oxide emissions, which have attracted the attention of environmentalists. The health of the environment is deteriorating day by day, and the ill effects of climate pollution are already being witnessed. Various pandemics, disease occurrences and the shrinking lifespan of human beings are directly or indirectly associated with climatic conditions; even international agencies [2] are addressing the correlation of various health hazards with climate safety. Moreover, the World Economic Forum report recently addressed the catastrophic Environmental Performance Index (EPI) of many developing countries compared with the global scenario [3]. The feed fermentation inside ruminants leads to significant dietary losses (6–10% gross energy loss and around 8–14% digestible energy loss), which results in higher input rearing costs [4].

Phytochemicals have strong antimicrobial properties and the potential to modify rumen fermentation to reduce enteric methane emission from ruminants [5]. The antimicrobial effect of essential oils suggests their relevance as anti-methanogenic feed additives, but it is challenging to maintain feed digestibility due to their general inhibitory effects on rumen microbes. Tannins are polyphenolic compounds and have the ability to modulate various rumen functions, viz., feed digestion, methane production, protein degradation, fatty acid biohydrogenation, etc., depending on the dose of supplementation. Condensed tannins near neutral pH form complexes with dietary proteins, carbohydrates, alkaloids, nucleic acids and minerals in the rumen, resulting in suppressed feed digestibility and the release of H_2 [6]. Tannins, especially condensed tannins, have marked impacts on rumen microbial proliferation [7]. However, the intensity of the effect varies with the source, type and concentration of tannins in the fermentation media. They may detrimentally affect rumen fermentation and reduce the production performance of animals [8]; however, their strategic use could improve animal health, milk production and the fatty acid profile in the milk of ruminants with a reduction in enteric methane and ammonia emissions [9].

Ficus glomerata, a species of plant in the family *Moraceae*, is native to Australia and tropical Asia. The leaves of the plant are rich in tannins, especially condensed tannins [10]. Therefore, the present study is conducted to examine the graded dose response of *Ficus glomerata* leaf extract on rumen fermentation, methane production and fatty acid biohydrogenation under an in vitro fermentation system with buffalo (*Bubalus bubalis*) rumen inoculum.

2. Results and Discussion

The modulation of rumen fermentation by plant bioactive compounds with the aim of reducing enteric methane production and enhancing the efficiency of animal production is an important research area for animal scientists as well as environmentalists. Tannins are naturally occurring plant secondary compounds that exhibit diversified action in animal systems depending upon their source, type, concentration, and delivery system to animals, basal diet, frequency of feeding, etc. [11]. The leaves of *Ficus glomerata* (FG) were found to be rich in polyphenolic compounds, with a significant concentration of condensed tannins (Table 1), which have the potential to modulate rumen fermentation [10]. The chemical composition of the oat hay/concentrate mixture (Table 1) used as a substrate for in vitro studies exhibited values within the range for Indian feeds and fodders [12]. The total gas production (mL or mL/g dry matter (DM) incubated) was gradually reduced ($p < 0.01$) at the dose levels of FG-0.50 and FG-1.0 (Figure 1); however, it remained intermediary and comparable ($p > 0.05$) for FG-0.25 with the control and FG-0.50, indicating the effect of tannins on rumen microbial activity. Compared to the control, the methane concentration (%) in the head space gas, as well as total methane production (mL or mL/g DM incubated, or mL/g DM digested), were found to undergo a gradual reduction ($p < 0.01$) with increasing doses (0.25–1.0 mL) of FG extract. Tannins have inhibitory effects on rumen protozoa and methanogenic archaea depending on the source and dose levels [13,14]. *Ficus glomerata* leaf extracts are rich in condensed tannins (Table 1), which could have role in reducing the archaeal population, thereby lowering methane production (Figure 2). However, no effects on methane production by tannins have been reported by many researchers [15,16], probably due to the addition of lower dose levels in their experiments.

Table 1. Chemical composition (% dry matter basis) of mixed feed (oat hay/concentrate mixture) used as substrate during in vitro studies.

Attributes	Oat Hay	Concentrate Mixture [1]	*Ficus glomerata* Leaf Meal
Chemical composition			
Organic matter	93.30 ± 1.12	91.07 ± 1.01	88.50 ± 1.23
Crude protein	7.80 ± 0.56	19.50 ± 0.67	14.85 ± 0.35
Ether extract	3.07 ± 0.60	4.12 ± 0.45	4.54 ± 0.24
Total ash	6.70 ± 0.72	8.93 ± 0.63	11.50 ± 0.82
Neutral detergent fiber	58.40 ± 1.35	44.60 ± 1.23	38.70 ± 1.67
Acid detergent fiber	43.70 ± 1.73	21.06 ± 0.96	31.52 ± 1.11
Phenolic fractions			
Total phenolics [2]	-	-	16.92 ± 0.52
Total tannin phenolics [2]			15.71 ± 0.41
Non-tannin phenolics [2]			1.22 ± 0.06
Condensed tannins [3]			10.64 ± 0.33

[1] Composed of the following ingredients (% parts): wheat grains, 20; maize grains, 15; mustard cake, 20; groundnut cake, 10; wheat bran, 32; mineral mixture, 2.0; and common salt, 1.0. [2] As tannic acid equivalent. [3] As leucocyanidin equivalent.

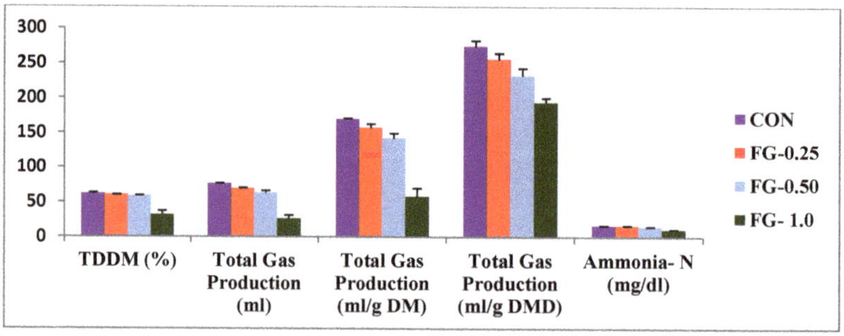

Figure 1. Effects of FG leaf extract supplementation in mixed feed substrate on truly degradable dry matter, total gas production and ammonia N.

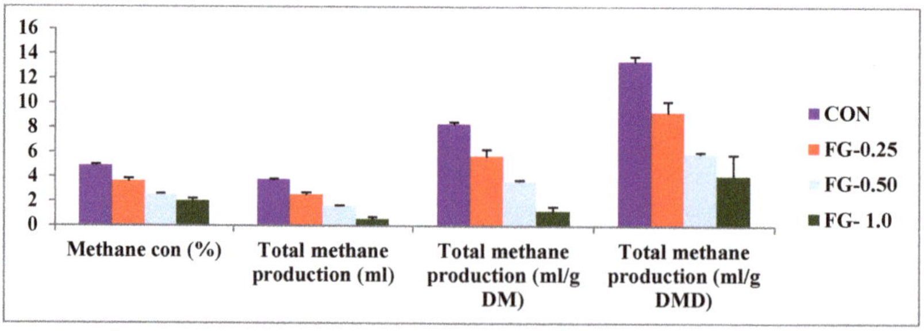

Figure 2. Effects of the FG leaf extract supplementation in mixed feed substrate on ruminal methanogenesis.

The reduced ($p < 0.05$) feed digestibility at higher levels (0.50–1.0 mL) of FG extract supplementation and comparative ($p > 0.05$) effects with the control (Table 2) at lower levels (FG-0.25) are suggestive of dose-responsive detrimental effects of tannins on fibrolytic microbes in the rumen. Reductions in anaerobic fungi, *Fibrobacter succinogenes* and *Ruminococcus flavefaciens* at 1 mg/mL of quebracho, mimosa and chestnut tannins were described [17]. Tannins at higher concentrations in diets remain free after binding with proteins and suppress fiber digestion by complexing with lignocellulose [18]. The reduction in gas production and methanogenesis at higher levels of FG extract (Figure 1) could also be attributed to reduced feed digestibility. The ammonia concentration decreased ($p < 0.05$) gradually in all the incubations in comparison to the control (Table 1). Tannins have an affinity for bacterial and plant proteins and reduce ruminal degradation [10,19], and thereby, ruminal ammonia production. However, ruminal protein protection and ammonia production depend on the type of tannins (hydrolysable or condensed tannins) [20]. The supplementation with FG leaf extract, which is rich in condensed tannins, might be responsible for the protection of substrate proteins, hence the reduced ammonia-N in the fermentation fluid (Figure 1). Supplementation with FG extract affected volatile fatty acid production in a dose-dependent manner. The molar proportion of acetate remained similar ($p > 0.05$) to that of the control in FG-0.25; however, it was reduced at higher doses (FG-0.50 and FG-1.0). Propionate production was enhanced ($p < 0.01$) in FG-0.25 but remained comparable ($p > 0.05$) in FG-0.50 with the adverse effect in FG-1.0 treatment. Butyrate production was comparable ($p > 0.05$) in all the treatments, except FG-1.0, where a reduction ($p < 0.01$) was observed. As the propionate production was enhanced ($p < 0.01$) in FG-0.25 and FG-0.50, the ratio of acetate to propionate was reduced; however, it remained comparable ($p > 0.05$) to the control in FG-1.0. The effects of tannins on rumen fermentation depend on the source, dose level as well as the substrate [5]. FG leaf extract is rich in condensed tannins (Table 1), which might have negative effects on fibrolytic rumen microbes at higher doses (FG-0.50 and FG-1.0), resulting a reduction in feed degradability and volatile fatty acid production [17]. While examining the effects of increasing doses of condensed tannin extract from *Cistus ladanifer* on in vitro ruminal fermentation and biohydrogenation, Guerreiro et al. [21] demonstrated a reduction in individual fatty acid production at all the studied doses. In the present study, the positive effect on propionate production at a lower dose level (FG-0.25) could be due to a shift in rumen fermentation via the modulation of rumen microbes and the reduction in methane production [9,13].

The modulation of rumen fermentation for reduced biohydrogenation of fatty acids is imperative for improving the quality of animal products [22]. Plant bioactive compounds, especially tannins, could modulate the biohydrogenation process to enhance mono- and poly-unsaturated fatty acids in milk and meat [14]; however, the effect varies with the type of tannins and their concentration [23]. *Ficus glomerata* leaves are rich in condensed tannins, and their extracts at graded levels of supplementation were found to modify ($p < 0.05$) various individual saturated fatty acids; however, the production of stearic acid (C18:0), a product of the rumen biohydrogenation process, was reduced ($p < 0.05$), irrespective of the concentration of FG extract (Table 3). Compared to the control, the concentration of *t*-vaccenic acid (C18:1), a precursor for conjugated linoleic acid (CLA) in animal products, was increased in all the FG extract-supplemented groups (Table 4). In an in vitro experiment by Carreño et al. [24] with four different tannin extracts (grape, quebracho, oak and chestnut), the inhibition of biohydrogenation with increased vaccenic acid concentration was described; however, a difference in their effects was delineated depending on the source and concentration. The supplementation of condensed tannins from *Acacia mearnsii* in a grass clover hay-based diet was demonstrated to enhance the intermediary products of biohydrogenation (cis-9, trans-11, cis-15 C18:3, trans-11, cis-15 C18:2 and trans-11 C18:1), suggesting the effectiveness of *Acacia mearnsii* CT for modulating biohydrogenation by removing potential microbes involved in BH. The tannin extracts of FG leaves, in the present study, might have modified the ruminal microbes, resulting in an increase (Figure 3) in *t*-vaccenic acid in the fermentation fluid.

Table 2. Effects of graded doses of tanniferous *Ficus glomerata* (FG) leaf extract on in vitro gas production, truly degradable dry matter, methanogenesis, volatile fatty acid production and fermentation characteristics of mixed feed substrate.

Attributes	Control	FG-0.25	FG-0.50	FG-1.0	SEM	*p* Value
Total gas production (mL)	76.55 ^{c} ± 0.47	69.32 ^{bc} ± 1.32	63.48 ^{b} ± 3.06	26.16 ^{a} ± 5.27	6.85	<0.001
Total gas production (mL/g DM)	169.52 ^{c} ± 1.14	157.07 ^{bc} ± 5.57	142.36 ^{b} ± 6.85	57.95 ^{a} ± 11.83	15.35	<0.001
Methane concentration (%)	4.88 ^{c} ± 0.12	3.60 ^{b} ± 0.23	2.53 ^{a} ± 0.06	2.0 ^{a} ± 0.21	0.41	<0.001
Total methane production (mL)	3.73 ^{d} ± 0.07	2.50 ^{c} ± 0.20	1.60 ^{b} ± 0.04	0.54 ^{a} ± 0.16	0.43	<0.001
Total methane production (mL/g DM)	8.27 ^{d} ± 0.18	5.67 ^{c} ± 0.56	3.61 ^{b} ± 0.09	1.19 ^{a} ± 0.36	0.96	<0.001
Total methane production (mL/g DMD)	13.36 ^{c} ± 0.46	9.22 ^{b} ± 0.91	5.89 ^{ab} ± 0.13	4.03 ^{a} ± 1.80	1.37	0.003
TDDM (%)	61.96 ^{c} ± 1.46	60.45 ^{c} ± 0.03	56.28 ^{b} ± 0.21	31.82 ^{a} ± 5.32	4.49	0.001
Ammonia N (mg/dL)	16.33 ^{c} ± 0.47	16.10 ^{b} ± 0.70	15.40 ^{b} ± 0.0	10.50 ^{a} ± 0.70	0.84	0.002
Acetate (mM/dL)	3.46 ^{c} ± 0.08	3.40 ^{c} ± 0.07	2.96 ^{b} ± 0.02	2.19 ^{a} ± 0.07	0.10	0.002
Propionate (mM/dL)	0.88 ^{b} ± 0.02	1.02 ^{c} ± 0.04	0.84 ^{b} ± 0.06	0.57 ^{a} ± 0.07	0.04	0.002
Butyrate (mM/dL)	0.25 ^{b} ± 0.01	0.23 ^{b} ± 0.02	0.24 ^{b} ± 0.01	0.15 ^{a} ± 0.01	0.02	0.001
A:P ratio	3.93 ^{c} ± 0.02	3.33 ^{a} ± 0.01	3.52 ^{b} ± 0.05	3.84 ^{c} ± 0.04	1.02	0.001

Control, FG-0.25, FG-0.50 and FG-1.0 are treatment groups with 0.0, 0.25, 0.50 and 1.0 mL of FG leaf extract/60 mL of buffered rumen fluid (BRF), respectively. Mean values bearing a, b, c and d superscripts in a row vary significantly ($p < 0.01$).

Table 3. In vitro fatty acid biohydrogenation (mg/incubation) profile upon incubation of mixed feed substrate with buffered rumen fluid supplemented with graded levels of tanniferous *Ficus glomerata* (FG) leaf extract.

Fatty Acids	Control	FG-0.25	FG-0.50	FG-1.0	SEM	*p* Value
		Saturated fatty acids (SFA)				
C4:0	36.75 ^{bc} ± 0.22	37.11 ^{c} ± 0.05	36.29 ^{b} ± 0.19	0.0 ^{a} ± 0.0	4.79	<0.001
C6:0	3.92 ^{c} ± 0.09	3.34 ^{b} ± 0.06	3.42 ^{b} ± 0.04	0.0 ^{a} ± 0.0	0.47	<0.001
C8:0	0.0 ^{a} ± 0.0	3.55 ^{b} ± 1.78	9.30 ^{c} ± 0.46	2.72 ^{ab} ± 0.18	1.09	0.001
C10:0	3.82 ^{b} ± 0.09	4.04 ^{c} ± 0.08	4.20 ^{c} ± 0.07	0.0 ^{a} ± 0.0	0.53	<0.001
C11:0	4.48 ^{c} ± 0.09	4.71 ^{c} ± 0.12	4.00 ^{b} ± 0.07	0.0 ^{a} ± 0.0	0.58	<0.001
C12:0	5.74 ^{a} ± 0.09	5.64 ^{a} ± 0.07	5.22 ^{a} ± 0.01	10.95 ^{b} ± 0.49	0.72	<0.001
C13:0	2.13 ^{a} ± 0.09	2.31 ^{a} ± 0.23	2.78 ^{a} ± 0.09	13.17 ^{b} ± 0.88	1.42	<0.001
C14:0	6.74 ^{a} ± 0.04	5.15 ^{a} ± 0.11	5.27 ^{a} ± 0.05	43.76 ^{b} ± 4.8	5.07	<0.001
C15:0	2.71 ^{a} ± 0.10	2.13 ^{a} ± 1.4	0.37 ^{a} ± 0.37	10.48 ^{b} ± 0.80	1.27	<0.001
C16:0	16.47 ^{d} ± 0.39	13.52 ^{c} ± 0.09	10.29 ^{b} ± 0.13	0.0 ^{a} ± 0.0	1.87	<0.001
C18:0	15.27 ^{b} ± 0.14	5.40 ^{b} ± 0.15	0.6.32 ^{c} ± 0.22	2.86 ^{a} ± 0.08	0.39	<0.001
Total SFA	88.05 ± 0.29	87.93 ± 0.52	87.49 ± 0.19	84.32 ± 3.64	0.89	0.53
C14:1	11.94 ^{d} ± 0.29	10.89 ^{c} ± 0.19	9.05 ^{b} ± 0.12	7.69 ^{a} ± 0.44	0.50	<0.001
Trans-vaccenic acid (C18:1)	0.0 ^{a} ± 0.0	2.89 ^{ab} ± 0.39	3.46 ^{ab} ± 0.29	7.99 ^{b} ± 3.49	1.16	0.063
Total unsaturated fatty acids	11.94 ± 0.29	13.07 ± 0.52	12.51 ± 0.19	15.68 ± 3.67	0.89	0.53

Control, FG-0.25, FG-0.50 and FG-1.0 are treatment groups with 0.0, 0.25, 0.50 and 1.0 mL of FG leaf extract/60 mL of buffered rumen fluid (BRF), respectively. Mean values bearing a, b, c and d superscripts in a row vary significantly ($p < 0.001$).

Table 4. Modulation (%) of rumen bio-hydrogenation of fatty acids, total gas production, methane production and truly degradable dry matter upon incubating graded doses of tanniferous FG leaf extract with mixed feed substrate.

Dose Incubated (mL/60 mL BRF)	SFA Inhibited	*t*-Vaccenic Acid (C18:1) Enhanced	Gas Production Inhibited	Methane Production Inhibited	TDDM Inhibited
0.25	0.14	100	9.44	32.98	2.01
0.50	0.64	100	17.07	57.10	8.73
1.0	4.24	100	65.83	85.52	48.64

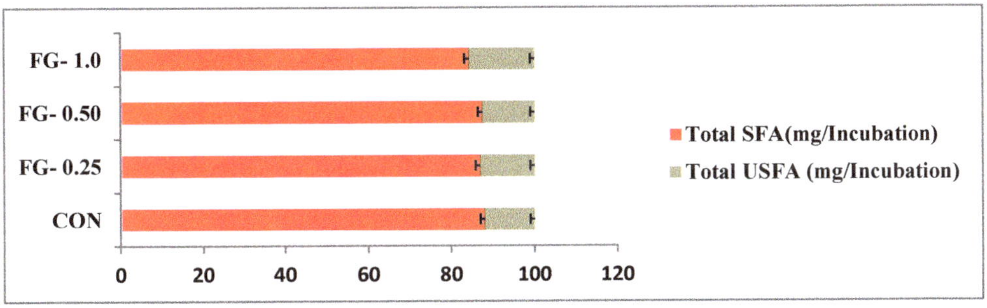

Figure 3. Effects of FG leaf extract supplementation with mixed feed on rumen biohydrogenation characteristics.

3. Materials and Methods

The experiment was conducted in the Division of Animal Nutrition and Feed Technology, ICAR—Central Institute for Research on Buffaloes (CIRB), Hisar, Haryana, India (29.1203_N, 75.8069_E). The guidelines of the Institutional Animal Ethics Committee of the CIRB were compiled (IAEC-CIRB/19-20/A/010 dated 5 August 2019) for the care of the rumen-fistulated animals and for rumen fluid collection.

3.1. Experimental Design and Substrate

A total mixed feed (500 mg) consisting of oat hay/concentrate mixture (1:1) was used as a substrate for this in vitro fermentation study. Tannin extracts (70% aqueous acetone extract) of *Ficus glomerata* (FG) leaves were prepared [25] and examined at four graded concentrations (0 mL, 0.25 mL, 0.50 mL and 1.0 mL per 60 mL of buffered rumen fluid) on the basis of preliminary laboratory screening. Three bottles per treatment per run with two incubation runs were prepared to obtain representative results.

Oats fodder was cultivated at ICAR—the Central Institute for Research on Buffaloes, Hisar, Haryana, India, harvested at the pre-bloom stage, oven dried at 65 °C for 48 h and ground to pass through 1.0 mm sieve for subsequent use. A concentrate mixture was prepared by mixing the following ingredients (% parts): wheat grains, 20; maize grains, 15; mustard cake, 20; groundnut cake, 10; wheat bran, 32; mineral mixture, 2.0; and common salt, 1.0.

3.2. Collection of Rumen Inoculum

Rumen liquor was collected from two rumen-fistulated Murrah buffalo steers (*Bubalus bubalis*) maintained on roughage-based diet at the animal farm of the institute. The mixed semisolid contents of the rumen digesta were collected manually from different locations at different depths in the early morning before offering feed and water to the animals and hand-squeezed to obtain the liquid portion of the rumen fluid. An equal volume of rumen fluid from both the fistulated animals was pooled to completely fill a 1.0 L pre-warmed oxygen-free thermos flask and brought to the laboratory. The rumen fluid was filtered through four layers of muslin cloth under continuous flow of CO_2 to use as a source of inoculums for the in vitro investigations.

3.3. *In Vitro Incubation*

Investigations were carried out using a control (CON) of total mixed feed as substrate and various treatment groups with different graded dose regimen (0 mL, 0.25 mL, 0.50 mL and 1.0 mL per 60 mL of buffered rumen fluid) of FG leaf extract. The in vitro incubations were carried out in 250 mL capacity ANKOM bottles fitted with radio-frequency (RF)-based pressure- and temperature-sensitive modules (ANKOM-RF Gas Production System). The day before the incubation, 500 mg mixed feed substrate was weighed into the bottles. On the day of in vitro incubation, the various concentrations of FG leaf extract were added to the respective bottles in duplicate. Incubation medium (60 mL) consisting of rumen liquor and in vitro buffer medium in a 1: 2 ratio [26] were added to the bottles under continuous gassing with CO_2 before being capped and returned to the incubator rack set at 39 °C and 85 oscillations/minutes for 24 h. Three bottles containing only incubation media without substrate were used as blanks for the experiment and three replicates for each treatment were maintained.

3.4. *Estimation of Fatty Acid Biohydrogenation*

After 24 h of fermentation, sample preparation for studying the in vitro fatty acid biohydrogenation of the substrate was carried out in accordance with the protocol of Mandal et al. [27], and finally, the sample was subjected to GC analysis. Incubated buffered rumen fluid (BRF) was initially prepared for esterification compounds, i.e., FAME (fatty acid methyl esters). For this, 20 µL of fermented BRF was added to Pyrex culture tubes (dimension, 16 × 125) and mixed with 0.35 mL of 10 N KOH and 2.65 mL of methanol. The contents were subjected to 1.5 h of incubation at 55 °C with 5 s of vigorous shaking at every 20 min interval. After incubation, the sample was subjected to cooling under running tap water, and 0.29 mL of 24 N H_2SO_4 was added with inverse mixing. The contents were again incubated and cooled in a similar manner. Finally, 1.5 mL n-hexane was mixed in the sample and vortexed for 5 min, and afterwards, subjected to centrifugation at 2500 rpm for 5 min. The acquired glassy transparent supernatant n-hexane layer containing FAME was subjected to GC analysis (Agilent GC system-8890, Agilent Technologies, Palo Alto, CA, USA), maintained at a 260 PSI inlet pressure, a 175 °C column temperature and a 260 °C detector temperature. The individual fatty acid concentration was calculated as follows:

$$\text{Individual fatty acid concentration (g/100 g of fatty acid methyl ester)} = \frac{Area\ of\ individual\ fatty\ acid}{Total\ area\ of\ fatty\ acid} \quad (1)$$

3.5. *Estimation of Gas and Methane Production*

After exactly 24 h of incubation, the recording was stopped and the total gas production (mol) in the bottles was recorded automatically through cumulative pressure and temperature using an ANKOM RF Gas Production System, and the total gas was determined as follows:

$$\text{Total gas production (mol)} = \frac{Pressure\ in\ ANKOM\ RF\ Gas\ Production\ bottle\ (KPa) * Head\ space\ volume\ of\ ANKOM\ RF\ Gas\ Production\ bottle\ (mL)}{Temperature\ of\ ANKOM\ RF\ Gas\ Production\ bottle\ (Kelvin)\ * Avogadro's\ number\ (8.3144)} \quad (2)$$

The methane concentration (%) in the head space gas was estimated after 24 h of incubation using a gas chromatograph (Nucon-5700, New Delhi, India) installed with a flame ionization detector (FID) and a column (Porapak 'Q'). The gas sample (200 µL), with the help of a 1000 µL graduated gas-tight micro-syringe (Hamilton, Switzerland), was taken from the bottles and injected into the injector port of the GC. A mixed gas (CH_4:CO_2: 50:50) (Centurion Scientific, New Delhi, India) was used as a standard for comparison. The temperatures of the injector, detector and oven of the GC were maintained at 140 °C, 200 °C and 70 °C, respectively, and the column pressure for the carrier gas, hydrogen, was set at 10 psi. The proportion of methane (%) in the total gas concentration was calculated as follows:

$$\text{Methane in head space } (\%) = \frac{[\textit{Area covered by the sample}] * 50}{\text{Area covered by standard of gas}} \quad (3)$$

The total methane production was calculated by multiplying the total gas production by the methane concentration.

3.6. *Volatile Fatty Acid (VFA) Estimation*

After the termination of incubation (24 h), 1 mL of the supernatant of each bottle's content was added to a micro-centrifuge tube containing 0.20 mL of 25% metaphosphoric acid. The mixture was allowed to stand for 2 h at room temperature and centrifuged at $5000\times\, g$ for 10 min to obtain a clear supernatant and stored at −20 °C for subsequent VFA determination. On the day of analysis, a sample of 1 µL was taken using a Hamilton syringe and injected manually into the gas chromatograph (NUCON-5700, Nucon Engineers, New Delhi, India), which was equipped with a glass column packed with chromosorb 101 and flame ionization detector (FID). The column pressures for hydrogen and zero-moisture air were 20 psi and 10 psi, respectively. The temperature of the column oven was increased from 170 °C to 230 °C at 3 °C/min and held for 7 min. The injector and detector temperature of the FID were set at 240 °C and 250 °C, respectively. Three blanks were included for the correction of VFA produced from the inoculum in each run, and two runs were executed for each sample.

3.7. *In Vitro Dry Matter Degradability and Ammonia Production*

The content of the each ANKOM bottle was transferred into a spoutless beaker via repeated washings with neutral detergent solution. After refluxing the contents for 1 h, the residue was recovered in pre-weighed filter crucibles (G1). After drying the crucibles to a constant weight, ashing was performed at 550 °C. The truly degradable dry matter (TDDM) was calculated as follows:

$$\text{TDDM } (\%) = \frac{(Dry\ weight\ of\ substrate\ incubated - Dry\ weight\ of\ residue\ leftover) * 100}{Dry\ weight\ of\ substrate\ incubated} \quad (4)$$

The ammonia nitrogen (NH_3-N) concentration was estimated in accordance with the Conway microdiffusion disk method [28].

3.8. *Chemical and Statistical Analysis*

Mixed feed substrates (oat hay/concentrate mixture) were chemically analyzed in accordance with the methods of the Association of Official Analytical Chemistry [29]. However, the fiber constituents were analyzed using the method of Van Soest et al. [30]. The extraction of dried *Ficus glomerata* leaves was performed using aqueous acetone (30:70), and the tannin fractions were analyzed [25]. The data obtained were subjected to analysis of variance (ANOVA) using SPSS 16.0 software [31], and the treatment means were ranked using Duncan's multiple range tests according to Snedecor and Cochran [32]. The mean differences were described via Tukey's HSD test, where the significance levels were $p \le 0.05$, which corresponds to 95% confidence, and $p \le 0.01$, which corresponds to 99% confidence.

4. Conclusions

Ficus glomerata leaf tannins have the ability to modulate rumen fermentation for reduced methanogenesis and fatty acid biohydrogenation in a total mixed ration. At all the experimental dose levels in this study, reductions in methane and ammonia production with enhanced *t*-vaccenic acid, a precursor of conjugated linoleic acid (CLA), were evidenced. As at higher levels (FG-0.50 and FG-1.0) of *Ficus glomerata* leaf extracts inclusion in in vitro fermentation medium, negatively affected feed digestibility and volatile fatty acid production, a lower dose (0.25 mL/60 mL or 4.17 mL/L) is suggested to achieve the desirable effects of a reduction in methane production and fatty acid biohydrogenation without affecting feed digestibility. However, an in vivo study would validate these results and determine its effectiveness for use as a feed supplement for the abatement of enteric methane emission and enhancing unsaturated fatty acids in animal products.

Author Contributions: Conceptualization: A.D.; data curation: R.K.S. and M.S.; methodology: R.K.S.; validation: M.S.; investigation: R.K.S. and A.D.; writing—original draft: R.K.S.; review and editing: A.D. All authors have read and agreed to the published version of the manuscript.

Funding: This study was funded by the IRC-approved project of ICAR-CIRB on the Development of a Feeding Module for Increasing Health-Promoting Fatty Acids in Milk and Reducing Methane Production in Buffalo (IXX08806).

Institutional Review Board Statement: All procedures involving animals met the guidelines of the "Committee for the Purpose of Control and Supervision on Experiments on Animals (CPCSEA)" issued by the Ministry of Environment, Forest and Climate Change, Government of India, with the approval (approval No. IAEC-CIRB/19-20/A/010 dated 5 August 2019) of the Institute Animal Ethics Committee (IAEC).

Informed Consent Statement: Not applicable.

Data Availability Statement: Not applicable.

Acknowledgments: The authors sincerely acknowledge the access to research facilities provided by the Director of ICAR–Central Institute for Research on Buffaloes, Hisar, and the Indian Council of Agricultural Research (ICAR), New Delhi, India.

Conflicts of Interest: The authors declare no potential conflict of interest relevant to this article.

References

1. Getabalew, M.; Alemneh, T.; Akeberegn, D. Methane production in ruminant animals: Implication for their impact on climate change. *Concepts Dairy Vet. Sci* **2019**, *2*, 204–211.
2. Pearce, W.; Holmberg, K.; Hellsten, I.; Nerlich, B. Climate change on Twitter: Topics, communities and conversations about the 2013 IPCC Working Group 1 report. *PLoS ONE* **2014**, *9*, e94785. [CrossRef] [PubMed]
3. Bhandari, M.P. Environmental performance and vulnerability to climate change: A case study of India, Nepal, Bangladesh and Pakistan. In *Climate Change and Disaster Risk Management*; Filho, W.L., Ed.; Springer: Berlin/Heidelberg, Germany, 2013; pp. 149–167.
4. Cottle, D.J.; Nolan, J.V.; Wiedemann, S.G. Ruminant enteric methane mitigation: A review. *Anim. Prod. Sci.* **2011**, *51*, 491–514. [CrossRef]
5. Nawab, A.; Li, G.; An, L.; Nawab, Y.; Zhao, Y.; Xiao, M.; Tang, S.; Sun, C. The potential effect of dietary tannins on enteric methane emission and ruminant production, as an alternative to antibiotic feed additives—A review. *Ann. Anim. Sci.* **2020**, *20*, 355–388. [CrossRef]
6. Schofield, P.; Mbugua, D.; Pell, A. Analysis of condensed tannins: A review. *Anim. Feed. Sci. Technol.* **2001**, *91*, 21–40. [CrossRef]
7. Waghorn, G. Beneficial and detrimental effects of dietary condensed tannins for sustainable sheep and goat production—Progress and challenges. *Anim. Feed. Sci. Technol.* **2008**, *147*, 116–139. [CrossRef]
8. Naumann, H.; Sepela, R.; Rezaire, A.; Masih, S.E.; Zeller, W.E.; Reinhardt, L.A.; Robe, J.T.; Sullivan, M.L.; Hagerman, A.E. Relationships between structures of condensed tannins from Texas legumes and methane production during in vitro rumen digestion. *Molecules* **2018**, *23*, 2123. [CrossRef] [PubMed]
9. Vasta, V.; Daghio, M.; Cappucci, A.; Buccioni, A.; Serra, A.; Viti, C.; Mele, M. Invited review: Plant polyphenols and rumen microbiota responsible for fatty acid biohydrogenation, fiber digestion, and methane emission: Experimental evidence and methodological approaches. *J. Dairy Sci.* **2019**, *102*, 3781–3804. [CrossRef]
10. Dey, A.; Dutta, N.; Sharma, K.; Pattanaik, A. Evaluation of condensed tannins from tropical tree leaves and its impact on in vitro nitrogen degradability of groundnut cake. *Anim. Nutr. Feed. Technol.* **2006**, *6*, 215–222.
11. Ashwin, K.; Pattanaik, A.K.; Howarth, G.S. Polyphenolic bioactives as an emerging group of nutraceuticals for promotion of gut health: A review. *Food Biosci.* **2021**, *44*, 101376. [CrossRef]
12. Ranjhan, S. Dairy feeding systems. In *Smallholder Dairying in the Tropics*; ILRI (International Livestock Research Institute): Nairobi, Kenya, 1999; pp. 117–132.
13. Bhatta, R.; Saravanan, M.; Baruah, L.; Prasad, C. Effects of graded levels of tannin-containing tropical tree leaves on in vitro rumen fermentation, total protozoa and methane production. *J. Appl. Microbiol.* **2015**, *118*, 557–564. [CrossRef] [PubMed]

14. Patra, A.K.; Saxena, J. Exploitation of dietary tannins to improve rumen metabolism and ruminant nutrition. *J. Sci. Food Agric.* **2011**, *91*, 24–37. [CrossRef] [PubMed]
15. Beauchemin, K.; McGinn, S.; Martinez, T.; McAllister, T. Use of condensed tannin extract from quebracho trees to reduce methane emissions from cattle. *J. Anim. Sci.* **2007**, *85*, 1990–1996. [CrossRef] [PubMed]
16. Min, B.; Pinchak, W.; Anderson, R.; Fulford, J.; Puchala, R. Effects of condensed tannins supplementation level on weight gain and in vitro and in vivo bloat precursors in steers grazing winter wheat. *J. Anim. Sci.* **2006**, *84*, 2546–2554. [CrossRef] [PubMed]
17. Denman, S.E.; McSweeney, C.S. Development of a real-time PCR assay for monitoring anaerobic fungal and cellulolytic bacterial populations within the rumen. *FEMS Microbiol. Ecol.* **2006**, *58*, 572–582. [CrossRef] [PubMed]
18. Barry, T.N.; Manley, T.R. Interrelationships between the concentrations of total condensed tannin, free condensed tannin and lignin in *Lotus* sp. and their possible consequences in ruminant nutrition. *J. Sci. Food Agric.* **1986**, *37*, 248–254. [CrossRef]
19. Frutos, P.; Hervas, G.; Giráldez, F.J.; Mantecón, A. Tannins and ruminant nutrition. *Span. J. Agric. Res.* **2004**, *2*, 191–202. [CrossRef]
20. Bhatta, R.; Uyeno, Y.; Tajima, K.; Takenaka, A.; Yabumoto, Y.; Nonaka, I.; Enishi, O.; Kurihara, M. Difference in the nature of tannins on in vitro ruminal methane and volatile fatty acid production and on methanogenic archaea and protozoal populations. *J. Dairy Sci.* **2009**, *92*, 5512–5522. [CrossRef]
21. Toral, P.G.; Monahan, F.; Hervás, G.; Frutos, P.; Moloney, A. Modulating ruminal lipid metabolism to improve the fatty acid composition of meat and milk. Challenges and opportunities. *Animal* **2018**, *12*, s272–s281. [CrossRef]
22. Hassan, F.-u.; Arshad, M.A.; Ebeid, H.M.; Rehman, M.S.-u.; Khan, M.S.; Shahid, S.; Yang, C. Phytogenic additives can modulate rumen microbiome to mediate fermentation kinetics and methanogenesis through exploiting diet–microbe interaction. *Front. Vet. Sci.* **2020**, *7*, 575801. [CrossRef]
23. Carreño, D.; Hervás, G.; Toral, P.G.; Belenguer, A.; Frutos, P. Ability of different types and doses of tannin extracts to modulate in vitro ruminal biohydrogenation in sheep. *Anim. Feed. Sci. Technol.* **2015**, *202*, 42–51. [CrossRef]
24. Guerreiro, O.; Alves, S.P.; Costa, M.; Duarte, M.F.; Jerónimo, E.; Bessa, R.J.B. Effects of Increasing Doses of Condensed Tannins Extract from *Cistus Ladanifer* L. on In Vitro Ruminal Fermentation and Biohydrogenation. *Animals* **2021**, *11*, 761. [CrossRef]
25. Makkar, H.P. *Quantification of Tannins in Tree and Shrub Foliage: A Laboratory Manual*; Springer Science & Business Media: Dordrecht, The Netherlands, 2003.
26. Menke, K.; Raab, L.; Salewski, A.; Steingass, H.; Fritz, D.; Schneider, W. The estimation of the digestibility and metabolizable energy content of ruminant feedingstuffs from the gas production when they are incubated with rumen liquor in vitro. *J. Agric. Sci.* **1979**, *93*, 217–222. [CrossRef]
27. Mandal, G.; Roy, A.; Patra, A. Effects of feeding plant additives rich in saponins and essential oils on the performance, carcass traits and conjugated linoleic acid concentrations in muscle and adipose tissues of Black Bengal goats. *Anim. Feed. Sci. Technol.* **2014**, *197*, 76–84. [CrossRef]
28. Conway, E. Ammonia. General Method. In *Microdiffusion Analysis and Volumetric Error*; Crosby Lockwood and Son Ltd.: London, UK, 1962; pp. 98–100.
29. Association of Official Analytical Chemistry (AOAC). *Official Methods of Analysis*, 18th ed.; AOAC International: Gaithersburg, MD, USA, 2007.
30. Van Soest, P.J.; Robertson, J.B.; Lewis, B.A. Methods for dietary fiber, neutral detergent fiber, and nonstarch polysaccharides in relation to animal nutrition. *J. Dairy Sci.* **1991**, *74*, 3583–3597. [CrossRef]
31. SPSS. *Statistical Packages for Social Sciences*, version 17.0; SPSS Inc.: Chicago, IL, USA, 2008.
32. Snedecor, G.; Cochran, W. *Statistical Methods*, 8th ed.; East West Press Pvt. Ltd.: New Delhi, India, 1994; p. 313.

methane

MDPI

Review

Measuring Livestock CH_4 Emissions with the Laser Methane Detector: A Review

Diana Sorg

German Environment Agency (Umweltbundesamt), Wörlitzer Platz 1, 06844 Dessau Rosslau, Germany; diana.sorg@uba.de

Abstract: The handheld, portable laser methane detector (LMD) was developed to detect gas leaks in industry from a safe distance. Since 2009, it has also been used to measure the methane (CH_4) concentration in the breath of cattle, sheep, and goats to quantify their CH_4 emissions. As there is no consensus on a uniform measurement and data-analysis protocol with the LMD, this article discusses important aspects of the measurement, the data analysis, and the applications of the LMD based on the literature. These aspects, such as the distance to the animal or the activity of the animals, should be fixed for all measurements of an experiment, and if this is not possible, they should at least be documented and considered as fixed effects in the statistical analysis. Important steps in data processing are thorough quality control and reduction in records to a single point measurement or "phenotype" for later analysis. The LMD can be used to rank animals according to their CH_4 breath concentration and to compare average CH_4 production at the group level. This makes it suitable for genetic and nutritional studies and for characterising different breeds and husbandry systems. The limitations are the lower accuracy compared to other methods, as only CH_4 concentration and not flux can be measured, and the high amount of work required for the measurement. However, due to its flexibility and non-invasiveness, the LMD can be an alternative in environments where other methods are not suitable or a complement to other methods. It would improve the applicability of the LMD method if there were a common protocol for measurement and data analysis developed jointly by a group of researchers.

Keywords: methane emission; methane measurement; measurement protocol; data processing; on-farm technique; method comparison

1. Introduction

Ruminants produce methane (CH_4) in their rumen and hindgut through enteric fermentation. This leads to losses of up to 12% of the gross energy intake and is a major source of greenhouse gas emissions [1]. Options to reduce CH_4 emissions from livestock include feeding strategies, feed supplements, and selective breeding [2–4]. Several methods have been developed to quantify CH_4 emissions from ruminants, such as the open-circuit respiration chamber (RC), which is very accurate and considered the "gold standard" for CH_4 measurements in ruminants [5]. However, when CH_4 emissions are to be measured on-farm, other methods such as the GreenFeed (GF) breath-analyser station (C-Lock Inc., Rapid City, SD, USA; [6]), non-dispersive infrared (NDIR)/Fourier-transformed infrared (FTIR) breath analysers ("sniffers") installed in feed bins [7,8], or the sulphur hexafluoride (SF_6) tracer gas technique [9] are used. All these methods have their own advantages and disadvantages in terms of purchase and running costs, labour, repeatability, behaviour change, and throughput [10] (Table 1), and they meet different requirements [11] (pp. 34–35). Another on-farm technique is the portable laser methane detector (LMD), which has comparatively low purchase and running costs and results in only low-to-moderate behavioural changes of the animals but requires relatively high labour resources and has a moderate throughput in terms of the number of records per time [10].

Table 1. Comparison of methods for measuring methane output by individual ruminants (adapted from [10] (p. 4)).

Method	Purchase Cost	Running Cost	Labour	Repeatability	Behaviour Alteration	Throughput
Respiration chamber	High	High	High	High	High	Low
SF_6 tracer gas technique	Medium	High	High	Medium	Medium	Medium
Breath analysers ("sniffers")	Low	Low	Low	Medium	None	High
GreenFeed	Medium	Medium	Low	Medium	Low	Medium
Laser methane detector	Low	Low	High	Low	Low–medium	Medium

The LMD was originally used to detect gas leaks from a safe distance in gas transmission networks, landfills, and other areas with CH_4 leakage risk [12]. In recent years, it has also been used to detect CH_4 concentrations in exhaled air from animals. Since the first known application for this purpose in 2009 by Chagunda et al. [12], researchers have further developed and evaluated the measurement, refined the analysis of data obtained with the LMD, and applied the LMD in studies on genetic analyses [13–15], on nutrition and feed efficiency [16–18], on the physiological status of animals [15], and to characterize different husbandry systems [19]. So far, measurements with the LMD have been carried out mostly on dairy cows, but also on sheep, beef cattle, and goats. A list of studies with the LMD used in this article can be found in Appendix A (Table A1).

There is no consensus on a uniform measurement and evaluation protocol with the LMD, so the results are poorly comparable. Therefore, this article aimed to provide an overview of relevant studies using the LMD and to discuss which aspects of the protocol for measuring and analysing LMD data have already been well studied, which could and should be standardised, and where further studies are still needed. An outlook will also be given as to which applications the method is suitable for and where, in the author's view, there are significant limitations.

2. The Laser Methane Detector

The measurement with the LMD (Figure 1a) is based on infrared absorption spectroscopy: it uses a semiconductor laser as a collimated excitation source and employs the second harmonic detection of wavelength-modulation spectroscopy for the measurement [12]. A visible guiding laser (Class 3 R laser, 532 nm) helps to direct the invisible measuring laser (Class 1 laser, 1653 nm) to the desired target. The integrated CH_4 concentration between the LMD and the target is measured by detecting a fraction of the diffusely reflected laser beam [20]. The measured value is expressed as CH_4 column density (ppm $\times$ m), i.e., a cumulative CH_4 concentration along the laser path or the average CH_4 concentration (ppm) multiplied by the length of the path (m) [21]. The LMD measures CH_4 in the range of 1 to 50,000 ppm $\times$ m (up to 5 vol-%) with an accuracy of $\pm$10%, and can be used from a distance between 0.5 and 30 m and in a temperature range from -17 to +50 °C. It autocalibrates via an internal reference cell [22]. The LMD shows the data in real time on its display and optionally issues an acoustic and visual alarm if a certain threshold is exceeded. Data can be stored in a csv file on a wirelessly connected Android device running the GasViewer app [23]. This can be, for example, a mobile phone (smartphone) worn in an armband sleeve so that one person can operate the LMD and the app at the same time. It has been originally designed to detect CH_4 from gas leaks in mining, the petrochemical industry, and landfills. The studies cited here used a "LaserMethane mini-g," a similar model or a previous model of the same series of LMD from the same manufacturer (Tokyo Gas Engineering Solutions, Tokyo, Japan).

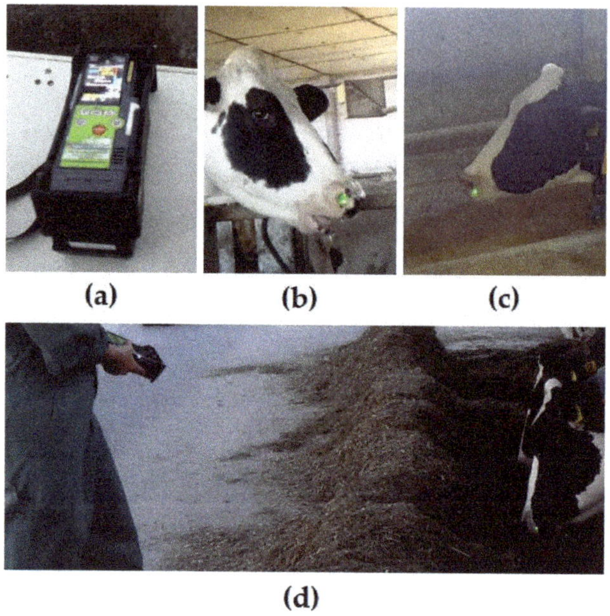

Figure 1. (**a**) A laser methane detector (LMD, Tokyo Gas Engineering Solutions, Tokyo, Japan). (**b**,**c**) cows in different positions with the visible guiding laser pointed at their nostrils. (**d**) measurement with the LMD (source: D. Sorg).

3. Aspects of the Measurement Protocol

CH_4 is transported from the rumen and lower intestine via the blood into the lungs and exhaled with the breath. The gas excreted directly from the rumen (eructation) is first inhaled into the lungs and then exhaled again with each respiratory cycle [24]. Only 2% to 3% of the CH_4 produced by the animal is released via the flatus [24,25]. Therefore, the LMD is aimed at the area around the animal's nostrils (Figure 1b,c), which is the main point source of emitted CH_4. Normally, an operator holds the LMD by hand (Figure 1d) and follows the animal's head movements, but it is also possible to mount the LMD firmly, e.g., on a tripod. In such a setup, it is necessary to fix the animal's head in one position to ensure a steady and uninterrupted measurement [26].

Chagunda et al. [12] were the first to use the LMD to record the CH_4 concentration in the breath of dairy cows [27]. They used the LMD from a distance of 3 m from the cow and took recordings at the nostrils for 15–25 s at a time. Since then, the LMD has been used in several studies to measure CH_4 breath concentrations in animals (Table A1). The measurement protocols differed mainly in terms of distance to the animal, duration of a single recording, the measurement interval during a single recording (i.e., the number of CH_4 values per time), and the number of repeats per animal (Table 2) but also in terms of time of day, animal activity, pointing angle, location, LMD operator, and the number of LMD devices used.

Table 2. Selected, quantitative aspects of the measurement protocol for the laser methane detector.

Variable	Minimum Observed	Reference (Example)	Maximum Observed	Reference (Example)
Distance to animal	1 m	[28]	3 m	[12]
Duration of recording	15 s	[12]	10 min	[29]
Measurement interval	0.1 s	[26]	5 s	[30]
Repeats per day	1	[14]	6	[31]
Repeats per animal	1	[14]	72 (cow); 63 (goat)	[26,32]
Consecutive days	1	[14]	10	[30]

3.1. Distance to the Animal

The evident advantages of a distance of 1 m are the direct conversion of the unit of measurement "ppm × m" to ppm (i.e., dividing the values by 1 m); the lower accumulation of background CH_4, which is added with every metre of distance along the laser path; and a lower influence of environmental factors such as air movements. It is also easier to track the target from a short distance, especially for goats, which are more mobile than cattle or sheep [26]. However, if such a short distance is not possible, e.g., if the natural behaviour of the animal is too much affected [31], a distance of up to 3 m seems acceptable. This distance should then be fixed for all measurements in an experiment, and a well-ventilated but windless environment is more important than for smaller distances.

3.2. Duration of Recording

In most studies, single LMD recordings were 3–5 min long. Shorter times are not recommended because an eructation event can be expected every one to three minutes [33,34]. Shorter recordings may, by chance, lead to lower mean CH_4 values if no eructation event is recorded or to significantly higher mean values if this is the case. Longer recordings have not been well studied. Sorg et al. [34] found no significant ($p > 0.05$) change in mean CH_4 values for recordings up to 9 min long compared to shorter recordings, and Doran [29] found no difference between mean CH_4 concentrations in the first and second half of a 10-min recording.

3.3. Measurement Interval

Most studies used a measurement interval of 0.5 s (i.e., two CH_4 values per second, the default setting of the LMD) [18,35,36]. Roessler et al. [26] recommended the use of 0.1 s—the shortest possible measurement interval—after showing that reducing data points to simulate a 1 or 4 s measurement interval resulted in increasing deviation of mean CH_4 values from the mean of the full data set. Since it is only a question of storage space and not affecting the amount of work involved in analysing the data, the shortest possible recording interval should be chosen.

3.4. Total Number of Repeats per Animal

Reported numbers of measurements per animal range from 1 [14] to 72 per cow [32] and 63 per goat [26]. When deciding on the appropriate number of replicates per animal, there is a trade-off between workload and invasiveness on the one hand and the quality of the data on the other. Studies with a small number of animals often use many replicates per animal, e.g., [17,26,30]. On the other hand, a study with a large number of animals (622 dairy cows) used one to three repeats per animal to calculate estimates of heritability (h^2) for CH_4 phenotypes [14]. So, it depends on the aim of the study and the available number of animals how many replicates per animal are feasible and appropriate. As with all measurement methods, increasing the number of measurement replicates is highly advisable, as this effectively reduces noise in the data [37] (p. 140), cited in [35].

3.5. Number of Consecutive Days per Measurement

Most LMD studies have used 3 consecutive days, e.g., [12,14,38], but up to 10 days have also been reported [30]. The former is similar to the number of days commonly used for RC measurements, but less than the protocol recommended for GF. For RC, a one-day measurement provides a reliable estimate of total CH_4 production within that day. However, in many studies, two to three days are used to reduce the variation in emissions caused by variations in dry matter intake (DMI). For GF, 20–50 spot samples are recommended, which would require 7–17 days of measurement if the animals visit the facility three times a day [39]. LMD measurements are more influenced by meteorological conditions and animal activity than GF measurements, and sampling over a few minutes is far less accurate than measuring over 24 h in a RC. Therefore, the number of consecutive days of LMD measurement is another important source of variation, and, so far, it is not clear what number of days gives the most reliable average value, but more than three is strongly recommended. However, the effect of measuring on several consecutive days should not be confused with repeating a round of measurement days some time later under different conditions. In the study of Mapfumo et al. [32], for example, the effect of different seasons on CH_4 emission was investigated in two measurement periods of six days each, three months apart. It should also be considered that the physiological state of the animals (lactation, growth, and performance) changes over such a long period of time, which leads to additional variation.

3.6. Time of Day

The time of day can have a significant influence on the measured concentrations [29,34,38], as CH_4 production in ruminants follows a diurnal pattern that is influenced by the timing of feeding and the times of rumination [36,40,41]. Therefore, it is recommended to perform all measurements in an experiment at approximately the same time of day and at the same distance from feeding to increase comparability or to include the time interval from feeding to the CH_4 recording as a fixed effect in the statistical model as described by Pinto et al. [19].

3.7. Animal Activity

The LMD has been applied on animals during different activities (eating, drinking, ruminating, lying, standing idle, and sleeping), being restrained or free to move in a barn or on pasture [27]. Considerable variation in CH_4 concentrations has been found during different activities; however, the ranking of the animals was not consistent between studies [12,26,40]. Drinking was among the activities with the highest CH_4 concentrations, presumably because the influx of water into the rumen triggered the flux of accumulated gas [36]. In a study with goats, neither activity nor restraint as opposed to free roaming had any effect on recorded CH_4 concentrations [31]. If possible, animal activity should be standardized within one experiment. If this is not possible, e.g., with free-roaming or grazing animals, it should be attempted to repeatedly record CH_4 during different activities for each animal and analyse the data for each activity separately.

3.8. Pointing Angle

This aspect has not yet been studied intensively, but, for purely practical considerations, it is of some importance for the measurement protocol. When measuring free-ranging animals, it is not always possible to aim exactly at the nostrils from the front, e.g., when the animal is facing a wall. If the laser beam of the LMD passes through a larger part of the exhaled CH_4 plume, the measured CH_4 values are higher. This could be the case if the laser beam is directed at the nostril at an angle corresponding to the direction of the exhaled air, as opposed to an angle perpendicular to the exhaled air. Sorg et al. [34] analysed the difference between pointing at cows in a tie-stall barn from the front or from the side in two different data sets and found that, on average, significantly ($p < 0.05$) higher CH_4 levels were recorded from the side (82 and 101 ppm $\times$ m) than from the front (73 and

93 ppm × m; $p < 0.05$). If the pointing angle varies during one experiment, this should be documented for each recording and considered in the statistical analysis.

3.9. Location

LMD measurements were performed at different locations inside buildings: in RC, e.g., [18,36]; in free-stalls, e.g., [12,14]; in a tie-stall [17]; in the milking parlour (according to the author's experience); a weighing facility for sheep [15]; and in individual pens, e.g., [38,42]. A few studies have been conducted on pasture [31,32,43] or in other half or full outdoor locations [19]. Apart from animal-related factors such as feed intake, the construction of the enclosure, the ventilation rate, and the presence of other animals seem to affect measured CH_4 levels, considering that the results from different experimental setups vary considerably, and significantly lower ($p < 0.05$) CH_4 concentrations were recorded outdoors than indoors [19]. Meteorological factors such as relative humidity and barometric pressure have been shown to have a weak but positive relationship with outdoor CH_4 concentrations, presumably by slowing the upward movement of CH_4 in the air [36]. Wind speed had a negative relationship with CH_4 concentration [36], probably due to a greater dilution of CH_4 in the ambient air. It can be assumed that the same mechanisms apply to CH_4 concentrations measured indoors. For this reason, Mühlbach et al. [14] recorded the wind speed in the barn during their LMD measurements and included it in their statistical model for genetic parameters for CH_4. Similarly, Reintke et al. [15] and Pinto et al. [19] accounted for temperature and humidity. Van Wyngaard et al. [43] even concluded that the LMD was not a practical tool for use with grazing animals in their region (Southern Cape, South Africa) and that weather conditions did not allow for successful use. After all these experiences with the LMD in different locations, it seems favourable to use a windless, but well-ventilated, environment with low and stable background CH_4 concentrations for measurements with the LMD. Furthermore, Chagunda [44] emphasises the need to include statistics on environmental conditions in the analysis of the data.

3.10. Operator

Most reports do not mention whether different LMD devices or different operators were used in a study. Chagunda et al. [12] found no significant operator effect in their statistical analyses. One study compared three operators who randomly alternated between using one of three LMD devices and jointly collected a dataset of a total of 520 LMD profiles from dairy cows in a free-stall barn [34]. As a result, one individual recorded significantly ($p < 0.01$) higher average CH_4 levels (97 ppm × m) than the other two individuals (86 and 87 ppm × m). Different handling of the LMD (e.g., precision in pointing at the animal's nostrils) and body size of the operators (i.e., a different vertical pointing angle) may have been the determining factors here. Mühlbach et al. [14] therefore included the operator of the LMD in the statistical model in their analyses of LMD data that had been collected jointly by five different operators. If more than one operator records LMD data during one experiment, the individual should be documented for each recording and considered in the statistical analysis.

3.11. Device

Previously, it was shown that two LMD devices agreed well when recording CH_4 concentrations in the air of an RC or barn in parallel [40]. However, in the same study as mentioned above [34], one LMD device recorded higher CH_4 values at the cow's nostrils (101 ppm × m) than the other two devices (85 and 86 ppm × m; $p < 0.001$), regardless of the operator. Interestingly, the unit with the deviating values had been purchased later than the other two, which had been delivered at the same time, although all three units were from the same supplier and of the same model. It can only be speculated whether this deviation is due to the date of manufacture, a different calibration, or a drift in the hardware conditions of the auto-calibration cell in the unit. As with the operator and the wind speed, Mühlbach et al. [14] included this as a fixed effect in the statistical model for the estimation of genetic parameters for CH_4. If more than one LMD is to be used for a trial, the LMD unit number should be documented for each recording and considered in the statistical analysis.

3.12. Animal Welfare

To the author's knowledge, the effects of the LMD on animal welfare have not been studied so far. As it is a minimally invasive procedure and the handling of the animals does not differ from routine procedures (e.g., fixing in headgates or crates), it can be assumed that there are no animal-welfare issues related to the method itself. According to the instruction manual [21] (p. V), care must be taken to ensure that the visible guiding laser (and with it the invisible measuring laser) is cast away from the animal's eyes at all times to avoid harmful exposure and injury.

In summary, a simple measurement protocol for researchers who want to get started with the LMD method and who do not have much experience with its handling could be as follows: Measure for 3 min or longer from a distance of 1 m, facing a restrained animal from the front. The measurement should be taken in a windless but well-ventilated building for three or more consecutive days at the same time of day, during the same activity of the animal and at the same distance from the last feeding.

4. Steps in LMD Data Processing

The data generated by the LMD make a list of CH_4 values accompanied by a unique date and time stamp, a value for the quality of the reflection of the laser beam, and optionally an input of the GPS location (if this function is activated on the connected Android device). A single measurement consisting of a time series of CH_4 values belonging to a single animal can also be called a "profile" [14,26,35] and is stored in the GasViewer app as a single file. Such a profile consists of regular peaks and troughs representing the inhalation and exhalation of the respiratory cycle [35] (Figure 2). When an eructation event occurs, the individual peaks are much higher and can reach maxima that are multiples of the respiratory peaks. Usually, one or more of these eructation events are recorded in one profile.

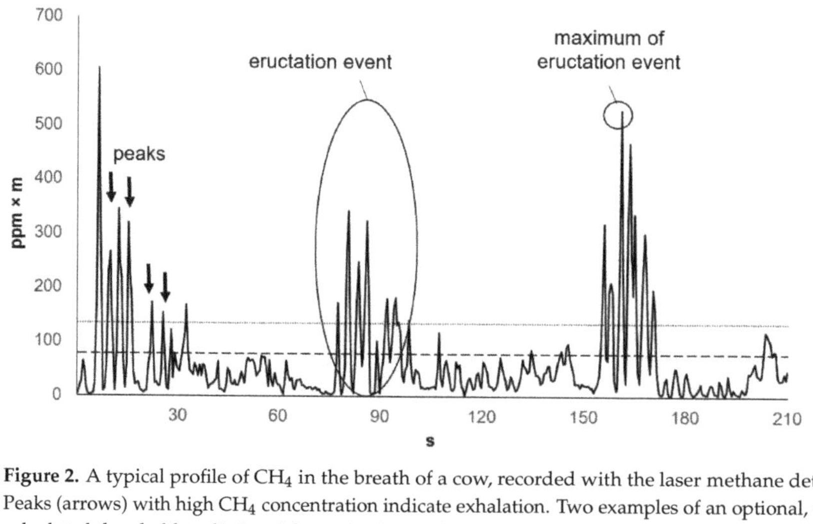

Figure 2. A typical profile of CH_4 in the breath of a cow, recorded with the laser methane detector. Peaks (arrows) with high CH_4 concentration indicate exhalation. Two examples of an optional, easily calculated threshold to distinguish respiration and eructation values are shown by the dotted line (boxplot method [35]) and the dashed line (one standard deviation from the mean of all values in the profile [28]).

In order to process the data from several profiles together, these files must be transferred to a computer and combined into one file. In doing so, the CH_4 values from each file should be given an individual number belonging to that file/profile so that they can be distinguished from each other during later analysis. This can be done manually, or, to facilitate the processing of a large number of files and minimise errors, with an automated, self-written computer script. Once a single file with all values is available, extensive processing of the data is required, which includes—but is not limited to—the following optional steps:

4.1. Exclusion of Profiles

A deletion of entire recordings was described in one study, where profiles were not categorisable into one of two fitted normal distributions (one for eructation and one for respiration; see step: "Separation of eructation from respiration") [18]. A more intuitive approach would be to plot all profiles and visually inspect the plots for any distinct deviation from the natural pattern of peaks and troughs. This seems only practical with a small data set and should be done very carefully in order to not introduce any unnecessary bias into the data.

4.2. Exclusion of Single Data Points

Some studies mention the deletion of single data points. One reason for this is unusually high CH_4 values, which are likely caused by errors in the reflectance of the laser beam [19,38]. These implausibly high values, e.g., above 2500 ppm $\times$ m (according to the author's experience) or more, depending on the experimental conditions, may occur when the recorded value for "reflection" is below 100 or when the visible pointing laser is reflected from a polished metal surface (according to the author's experience). For this reason, Niero et al. [42] deleted all values above three standard deviations from the overall mean. As with the exclusion of entire recordings, profiles can be visually inspected for isolated, very high peaks that are not part of the regularly occurring pattern of peaks but with the above limitations. Regardless of which variant is chosen, it should be ensured that unphysiologically high single values are detected and deleted. Another source of faulty data is the movement of the animal, which can lead to a time- and labour-consuming removal of data from times when the laser beam was off-target [18,31].

4.3. Accouting for Background CH_4 Concentration/Offset

The LMD has an internal "offset" function to subtract the background CH_4 from all subsequently recorded values. This was used in several studies, e.g., [12,31,45]. Other researchers used the minimum CH_4 concentration of each profile [15,19,38] or from all records [29] and subtracted it from all values in a profile. Rey et al. [46] assumed a fixed background concentration of 2 ppm and subtracted it from their measurements. However, the background CH_4 concentration can be highly variable and is unrelated to the day, time of day, or time within the measurement ([26], and according to the author's experience). Doran [29] concluded that the offset function of the LMD was not suitable in their experiment because there were many other animals and a dung heap nearby. So, the question remains whether this step is really necessary or whether it introduces bias by deliberately choosing a random value in a random location as a background.

4.4. Accounting for Distance

The recorded CH_4 can be directly converted from ppm $\times$ m to ppm by dividing it by the length of the laser path. This is useful when the distance is 1 m or less, or when the CH_4 concentration is relatively homogeneous over the entire distance between the LMD and the target, such as in an RC [36]. Rey et al. [46] assumed that the plume with the high CH_4 concentration in front of the animal was only 0.1 m long and therefore multiplied their measurements by 10. In many other studies, no such conversion or consideration of distance was included [12,30,31]. If the animals are to be ranked or different treatments are to be compared, absolute CH_4 values and accounting for distance are not necessary. Here it is only important to use the same experimental conditions for all animals and measurements to ensure comparability. It has also been shown that high CH_4 concentrations are only up to 0.4 m from the animal and decrease very quickly with increasing distance [34]. Dividing by a distance of 2 or 3 m would therefore artificially dilute the recorded CH_4 values. Sorg et al. [35] assumed that the additional low CH_4 background added by different distances (1 m, 2 m, and 2.5 m) at different measurement sites was negligible for the specific purpose of their study. Kobayashi, et al. [18] reported a variation in distance of 0.6 to 1.2 m in their experiment but did not describe any consideration of distance in the analysis of the data. However, if possible, the distance should be standardised within a study.

After a thorough quality control, the data can be transformed and the CH_4 concentrations of the individual animals should be combined into some kind of point measurement, also called a "trait" or "phenotype." Typical steps for creating these point measurements are described in the following Sections 4.5–4.9.

4.5. Transformation of the CH_4 Values

Within a profile there are many low (respiration, Figure 2) and few high CH_4 values (eructation, Figure 2), so the data do not follow a normal distribution. In some studies, therefore, all CH_4 values were $\log_e$-transformed (natural logarithm) to achieve a normal distribution and homogeneity of variances [42], or only the peak respiratory values ("mini-peaks," see "Separation of Eructation and Respiration") were $\log_e$-transformed [15]. Bruder et al. [33] found that when all values above the 95th (238 ppm) or 99th percentile (562 ppm) were deleted from their data set, a normal distribution could be achieved without log-transforming the data. Most other studies proceeded without any transformation of the individual values. It is not certain whether such a transformation is necessary when all values of a profile are later condensed into a single point measurement.

4.6. Separation of Peaks and Troughs

The regularly occurring CH_4 peaks in an LMD profile (Figure 2) represent the respiratory cycle [28,35]. These peaks in the time series of CH_4 values can be identified using an automated, self-written computer script or Excel macro (Microsoft Corporation, Redmond, WA, USA; [38]). Another simple identification method was used by Niero et al. [42]: They defined the last decile of the distribution of all CH_4 values as peak values. However, it is unclear whether this phenotype is comparable, in a physiological sense, to the one described above [35,38]. Niero et al. [42] concluded that repeatability and reproducibility were better for the average of the peaks than for all CH_4 values. However, the same authors acknowledged that the average of the peaks may not be suitable to distinguish between high and low CH_4 emitters due to its lower variability [42]. Sorg et al. [35] also found higher repeatability (between-cow variance/total variance) for the mean of peak values (0.24) than for the mean of all values (0.12) and subsequently used the former phenotype for their analyses. Rooke et al. [28] obtained similar results with the mean of all and the mean of the peak CH_4 values and subsequently used the mean of all values. The majority of the studies used the peak values. This fact and the physiological explanation that the peaks are the CH_4 concentration in the breath make it seem reasonable to at least calculate and analyse both phenotypes in parallel.

4.7. Separation of Eructation and Respiration

Series of very high CH_4 peaks in each LMD profile can be clearly attributed to eructation Rooke [28]. Several of these eructation events can be distinguished in a typical CH_4 profile (Figure 2). This is an advantage of the LMD over other CH_4 measurement methods: because the LMD has a very short measurement interval (up to 0.1 s), it can accurately resolve the fluctuations in CH_4 levels due to respiration and eructation and provide valuable information about physiological processes. Several methods have been developed to automatically separate the eructation values from the respiration values. An easily calculated threshold is one standard deviation (SD) above the mean of all values in a CH_4 profile [28,33,36]. Such a value appears physiologically reasonable (Figure 2) and takes into account the individual values of exhaled and background CH_4. Sorg et al. [35] used the definition from the boxplot method [47,48] (Figure 2), which has similar advantages: Possible outliers (i.e., eructation) in a distribution of values (i.e., all values of a profile) are above the threshold (T) with

$$T = Q3 + (1.5 \times (Q3 - Q1)), \quad (1)$$

$Q3$ being the third quartile and $Q1$ the first quartile of all values. A more sophisticated approach was introduced by Ricci et al. [38], who fitted two normal distributions (one for eructation and one for respiration) to the pool of all values from all records and calculated the probability that each value belonged to one of these distributions. The threshold for respiration was a cumulative probability of 99%, and eructation events were defined as at least two consecutive eructation peaks [38]. Other studies that applied this threshold used a probability of 95% for respiration [15], of 10% for eructation [19], or set the threshold such that all values were placed in the category for which the values had a higher probability [18]. It is still not clear whether it is necessary to analyse respiration and eructation of CH_4 separately. According to Ricci et al. [38], this separation improved the ability of LMD outputs to discriminate between feeding treatments and the correlation with CH_4 outputs of a RC. Furthermore, Kobayashi et al. [18] found a higher correlation with the CH_4 of the RC when only the respiration CH_4 from the LMD measurements was used. This contrasts with the higher repeatability (between-cow variance/total variance) of the mean of all peaks (0.24) compared to the mean of only the eructation peaks (0.11) or the mean of the maxima of eructation events (0.17) [35] and the higher estimates of h^2 for the mean of all peaks (0.07–0.23) than for the mean of the maxima of eructation events (0.05–0.08) [14].

4.8. Reduction of Data to Point Measurements

For the statistical analysis of CH_4 data obtained with the LMD, it is necessary in most cases to reduce the time series of CH_4 values to a single point measurement (or trait or phenotype). So far, there is no consensus on the best phenotype. The mean of the peak values seems to be the most studied phenotype [12,13,17,26,38], but other measures that summarise all values [28,30,42,46], peak values [19,38,42], respiration or eructation values [18], respiration or eructation peak values and eructation events [14,15,19,35,38,46] have also been analysed (Table 3). So far, different phenotypes based on total, respiration, and eructation values have performed inconsistently and differently across studies in terms of CH_4 prediction accuracy [18] and usability for genetic [14,15] and nutritional analyses [38]. It is therefore recommended to analyse and present relevant phenotypes from respiration, eructation, and total CH_4 (Table 3) in parallel as long as there is no consensus on the best phenotype.

The selection of phenotypes should also be based on the previous studies with which one's own results are to be compared.

Table 3. Possible point measurements (traits and phenotypes) that can be derived from data obtained with the laser methane detector.

Category	Point Measurement	Explanation [1]
All values	Mean	Breath CH_4 concentration including re-inhaled and exhaled eructation and background CH_4 concentration during inhalation
	Number	Not meaningful – number is pre-defined by measuring interval (e.g., 2 per s)
	Maximum	Highest single CH_4 concentration
	Sum	Cumulative CH_4 concentration including background

Table 3. *Cont.*

Category	Point Measurement	Explanation [1]
Peaks	Mean	Breath CH_4 concentration including re-inhaled and exhaled eructation CH_4 without background
	Number	Proxy for breath frequency (not for CH_4 emission)
	Sum	Cumulative breath CH_4 concentration without background
Respiration peaks [2]	Mean	Breath CH_4 concentration without eructation
	Number	Breath frequency but without times of eructation—physiologically not meaningful
	Maximum	Highest non-eructation CH_4 peak
	Sum	Cumulative breath CH_4 concentration without eructation and background
	Time	Duration of respiration
	Percentage	Of respiration peaks from all peaks in a recording—can be used to validate physiological plausibility of the data
Eructation peaks	Mean	Breath CH_4 concentration from eructation only
	Number	Breath frequency during eructation—physiologically not meaningful
	Maximum	Highest single CH_4 concentration (same as for "all values")
	Sum	Cumulative breath CH_4 concentration from eructation only
	Time	Duration of eructation
	Percentage	Of eructation peaks from all peaks in a recording—can be used to validate physiological plausibility of the data
Eructation events	Mean of the maxima	Series of the highest CH_4 concentrations
	Number	Eructation frequency—not a real CH_4 phenotype but can be used to validate physiological plausibility of the data
	Maximum	Highest single CH_4 concentration (same as for "all values")
	Sum of the maxima	Cumulative breath CH_4 concentration from the series of the highest CH_4 concentrations

[1] according to the author's assessment and other studies [12–15,17–19,26,28,30,35,38,42,46]. [2] Separation into respiration and eructation can also be done with all values but would include background CH_4 during inhalation.

4.9. Estimation of Daily CH_4 Production

LMD measures a concentration, whereas most other widely used methods allow quantification (mainly the RC) or at least estimation of an animal's total CH_4 production in g/day [10]. Direct comparison or pooling of CH_4 data from LMD with data obtained using other techniques is therefore only useful if the aim is to rank animals rather than quantify their absolute CH_4 production. For this reason, there have been many attempts to establish equations to convert CH_4 concentrations measured with the LMD into total CH_4 production in g/day [27]. Chagunda et al. [12] estimated daily CH_4 production based on the CH_4 concentrations in the breath during different activities, the assumed duration of these activities in a day, and the average tidal volume during different activities. Similarly, Sorg et al. [34] estimated CH_4 production during a 5 min LMD measurement from average CH_4 breath concentration, individual number of CH_4 peaks (i.e., number of breathing cycles), and the individual tidal volume estimated from live weight [49] (p. 200). Both equations [12,34] produced values that were biologically meaningful, on average, but had very high standard deviations and some implausible low and high individual CH_4 production values [34]. In two studies with sheep [29] and goats [50], where similar equations were used, this deviation was much lower and allowed reasonable use of this phenotype for further analyses. Nevertheless, these equations are suitable to provide rough estimates of the average CH_4 production of a group of animals or a treatment group, but they are not suitable for the accurate quantification of an animal's individual CH_4 production.

Another approach to obtain data on total CH_4 production from LMD measurements is to use another technique as a reference and measure CH_4 with the LMD in parallel. In this way, a regression equation can be derived from the known CH_4 production (measured with the other technique) for LMD measurements. In a number of studies, the "gold standard," the RC, was used for this purpose. These results should be interpreted with caution, as the conditions in the RC are very different from those in a barn. In some of these studies, the LMD was applied while the animals were inside the RC [18,36]. CH_4 concentrations are higher in enclosed spaces than in well-ventilated barns or outdoors [19,40]. Therefore, LMD measurements in RC cannot be directly compared with those on-farm. Another approach is to measure on-farm CH_4 concentrations in the same environment as used for subsequent LMD studies and then move the animals to a RC for an estimate of daily CH_4 production, as described by Rooke et al. [28], Ricci et al. [38] and Kecman et al. [51]. It must be considered that the CH_4 output in the chamber may not be the same as the one on-farm. Animal behaviour, feed intake, and milk yield may be affected by the stress caused by handling, confinement, and the unfamiliar environment [10,52,53]. Therefore, equations derived from comparing LMD and RC are unlikely to reliably estimate on-farm CH_4 production. For this reason, total CH_4 production in g/day estimated with a GF system was used as a reference in the study by Sorg et al. [35]. The GF system was installed in the barn, and the cows had free access to it. Between these visits, recordings were made with the LMD in the same barn, so that both measurements were made in a similar environment and physiological status of the cows. The resulting regression equation (R^2 = 0.65) for daily CH_4 production from LMD measurements allowed a successful comparison of LMD and FTIR/NDIR sniffer measurements from different trials [35].

All the above-mentioned studies produced equations for an estimate of daily CH_4 production derived under the specific conditions in these experiments and adapted for a specific purpose. It is not recommended to apply these equations to LMD measurements from other experiments with different conditions without further validation or reference. Even with more research, it may not be possible to develop a universal equation suitable for all types of LMD measurements, as the conditions are very different between experiments. It remains a serious limitation to the method that only the concentration and not the flux of CH_4 can be measured.

Knowing the limitations of a simplified analysis, here is nevertheless a suggestion for a very basic protocol for the analysis of LMD data for researchers who want to start with the LMD method and do not yet have much experience with this type of data: Delete all CH_4 values above 2500 ppm $\times$ m and those with an intensity value < 100. Calculate the mean of all values for each profile. If you have access to mathematical computer software or programmes with customizable computing algorithms, determine peak values and calculate the mean of all peak values. Choose a simple threshold for respiration and eructation, e.g., the boxplot method or one standard deviation above the mean of each profile and calculate the mean of all respiration and eructation values for each profile.

5. Comparison with Other Techniques

Since the beginning of the use of the LMD in livestock, researchers have compared the LMD with other techniques for measuring CH_4 emissions from ruminants. It is important to distinguish between

1. Technical evaluations of the accuracy of the sensor and
2. Evaluations of the overall measurement and analysis protocol for the LMD.

For (1), Chagunda and Yan [45] and Sorg et al. [40] installed LMD units in RC and recorded ambient air CH_4 concentrations in the chamber by aiming the LMD at the gas outlets from inside the chamber. This was a comparison of the sensors of the RC and the LMD, not of the measurement principle of these techniques [11] (pp. 29–30). Both studies reported good agreement with CH_4 levels measured by the RC sensors, proving that the LMD is capable of accurately quantifying fluctuating and physiologically low CH_4 concentrations in the air inside buildings. This is an important prerequisite for further applications of the LMD on livestock, considering that it was originally developed for detecting very high CH_4 concentrations at gas leaks in mining, the petrochemical industry, and landfills.

For (2), the LMD was compared with RC, GF, and FTIR/NDIR sniffers, as well as with indirect methods using prediction equations for CH_4 production from proxies. One such prediction equation from feed intake and feed properties yielded lower average CH_4 values than the one estimated from LMD measurements [12]. Reported quantitative agreement ranges from none with CH_4 predicted from milk mid-infrared spectroscopy (MIR) and RC [17] to a Pearson correlation of 0.82 with RC expressed as total CH_4 production in g/day and kg milk [51]. A significant positive regression of LMD on RC CH_4 was also reported by Rooke et al. [28] ($R^2 = 0.27$, $p < 0.001$), or in the form of a positive and significant ($p < 0.05$) Pearson correlation by Chagunda et al. [36] (0.18 and 0.47 for sheep and dairy cows, respectively), Doran [29] (0.57), Rooke et al. [28] (0.53), Bruder et al. [33] (0.47), Brocklehurst et al. [54] (0.6), and Kobayashi et al. [18] (0.55). Ricci et al. [38] improved the relationship of LMD and RC with a model containing the LMD phenotypes "mean eructation time" or "maximum CH_4 concentration during respiration" and DMI. Kobayashi et al. [18] found the best correlation with RC CH_4 output using respiration only, in contrast to all values or eructation only.

In addition, Chagunda et al. [36] evaluated the ability of the LMD to detect high CH_4 levels (above the third quartile of all values) similarly to the RC sensor and reported a sensitivity of 95% and 94% and a specificity of 97% and 79% for cows and sheep, respectively.

When comparing methods, systematic differences between methods (means), any random differences (precision), and correlation between methods can only be assessed without residual error when two or more measurement replicates are performed per animal [10]. This was not always the case in these studies (e.g., when an overall average for LMD phenotypes was compared to an average CH_4 production from the RC). Therefore, simple correlations should be interpreted with caution and may not reflect true agreement or prove disagreement between methods.

Under on-farm conditions, a high repeated measures correlation (r_p = 0.66) was found using replicates of the LMD phenotype "mean of all peaks" and the total daily CH_4 production determined by GF [35]. In addition, a regression equation for total CH_4 production in g/d from the LMD was derived from this comparison and used to compare the LMD with an FTIR (r_p = 0.57) and an NDIR sniffer (r_p = 0.60) [35]. Although means and variances differed greatly between instruments, Sorg et al. [35] concluded that these values of the r_p implied minimal re-ranking of the animals in their experiment. Rey et al. [46] obtained an even higher r_p for the mean of all CH_4 values (0.98) and the number of eructation events (1.00) when they compared the LMD with an NDIR sniffer. This could be due to the fact that the sampling tube of the NDIR sensor was manually held to the cow's nostrils while measuring simultaneously with the LMD. In the study by Sorg et al. [35], it was placed in the feed container of an automatic milking system (AMS) and the measurement with the LMD was taken after the cow had left the AMS. A comparison with the SF_6 tracer gas technique in grazing cows was attempted by Van Wyngaard et al. [43], but the measurement was not successful due to weather conditions.

A systematic comparison of different on-farm techniques to measure CH_4 using bivariate models was conducted by researchers of the European Cooperation in Science and Technology (COST) Action METHAGENE (www.methagene.eu, accessed 23 December 2021, [10]). Data from only one study comparing the LMD with other techniques [35] were available to the group at that time. It would be helpful to systematically compare the LMD to the RC, the GF, sniffer sensors, the SF_6 tracer gas technique, CH_4 prediction equations, and possibly other new methods in one joint data set along the lines of this work. In this way, the agreement and important factors influencing it, the limitations, and the possibilities of the LMD could be better characterized, working towards a more standardized protocol for the measurement and the analysis of LMD data. From the above-mentioned studies, it becomes clear that the LMD has the potential to rank animals similarly to other techniques, even if the absolute CH_4 levels are different.

6. Applications

In several areas of ruminant research, there is a need to assess CH_4 emissions in order to develop mitigation measures or to obtain valuable information on the physiological and metabolic status of animals. Hence, the LMD has been successfully used for studies in various fields of animal science:

6.1. Genetics

In addition to the established and well-studied traits for animal performance, conformation, health and fertility, traits for feed efficiency and environmental impact have been studied for some time. Selective breeding for lower CH_4 emissions could make a valuable contribution to the set of mitigation strategies to achieve climate targets but requires data from a sufficient number of animals that are phenotyped and genotyped [55]. Such data must be obtained under on-farm conditions from a large sample of animals. Therefore, complex or invasive techniques with high accuracy such as the RC or SF_6 tracer gas methods are not suitable. The LMD is able to classify animals according to their CH_4 production and has therefore been used to calculate estimates of h^2 for several CH_4 phenotypes [13–15]. The results (0.01–0.23) are in the range of those obtained with other on-farm methods [56], although the standard errors are higher. The measurements with the LMD are more influenced by environmental conditions, resulting in higher variability of the recorded values. Furthermore, the number of animals that have been included in genetic evaluations based on LMD data so far is limited. With larger samples and a more standardised protocol for measurement and data analysis, estimates of h^2 are likely to improve.

6.2. Nutrition

Novel feeds and natural and synthetic feed additives are emerging CH_4-mitigation technologies with high commercialisation potential and a potentially large impact on CH_4 emissions in the future [4]. The effect of these feeds and feed additives needs to be validated in a large sample of animals under on-farm conditions. Among other methods, the LMD could be a valuable technique for their investigation. A few studies have successfully used the LMD to discriminate between different rations and feeding strategies at the group level [28,38,50]. Cameron et al. [57] measured the CH_4 production of dairy cows with the LMD and concluded that adding fresh grass or pasture to a total mixed ration reduced it by 17% and 39%, respectively, and could be a valuable tool to reduce CH_4 emissions. However, some researchers were unable to detect differences in CH_4 concentration or production measured with the LMD due to dietary components [19,26] despite their well-studied effect on methanogenesis [2].

Vrancken et al. [16] demonstrated the effectiveness of a novel feed additive to decrease CH_4 emissions from cows using LMD measurements. If such feed additives are to be used as mitigation measures in the future, their effect must be extensively demonstrated at the farm level. For this, the LMD could be a useful application.

6.3. Farming Systems and Breeds

Due to its relatively low cost, flexibility, and portability, the LMD can be used to systematically compare farms and farming systems in regions with limited access to research infrastructure or with animals on pasture. This was demonstrated by Pinto et al. [19], who were able to characterise access to pasture and different husbandry systems along an urban–rural gradient in India using CH_4 phenotypes obtained with the LMD. Grobler et al. [30] found breed differences ($p < 0.05$) in the CH_4 production of South African Jersey, Bonsmara, and Nguni cattle, while in the study by Mapfumo et al. [32] extensively reared African Boran and Nguni heifers showed no difference in CH_4 output per DMI. If the breed differences in CH_4 production are too small, the LMD may not be able to detect them.

6.4. Health and Metabolism

To the author's knowledge, the correlation of physiological blood parameters with CH_4 production measured with the LMD has only been investigated in one study so far. In the study by Reintke et al. [58], the blood concentrations of zinc, β-hydroxybutyrate, and unesterified fatty acids had a partly breed-specific significant ($p < 0.05$) influence on CH_4 concentrations.

For the applications described above, the LMD can be a complement to other established methods or an alternative in environments where more sophisticated and complex methods are not suitable. The LMD meets the requirement to measure variation over time (especially whole and consecutive lactation periods) in large numbers of animals without disrupting normal animal behaviour or farm management on commercial farms and limiting investment and maintenance costs [11] (pp. 34–35). It can be used for experiments where animals are to be categorized or ranked according to their CH_4 concentration—especially in genetic analyses—or where average values of CH_4 concentration or estimated CH_4 production at a group level are sufficient. For these purposes, large samples are recommended: many replicates per animal and/or many animals. A common protocol specific to each purpose would further improve the comparability of such studies and is also suggested by other researchers [27]. If possible, external validation with another on-farm method (e.g., GF, sniffer, or SF_6) or the newly developed "artificial cow" [59] should be attempted. It is not recommended to use the LMD if accurate values for absolute CH_4 production of individual animals are needed.

7. Conclusions

The LMD has several advantages: it is flexible, portable, easy to handle, and it does not require an external power source. It is therefore relatively inexpensive to use in many different experimental and commercial environments. Nevertheless, there are also substantial limitations: measurement with the LMD is labour- and time-intensive, and of all the established on-farm methods for CH_4 measurement, it is probably the least accurate, as it only measures concentrations, not quantities, and is strongly affected by environmental conditions. Absolute values for CH_4 production can only be calculated with some uncertainty and must be interpreted with caution. Nevertheless, it can be a complementary method to others or where alternatives are lacking or not applicable. The application of the LMD can be extended to various research questions and has already been demonstrated exemplarily for nutrition, physiology, and genetics.

From the first exploratory application of the LMD in 2009 to today, the research community has come a long way, and many questions have been answered about what the LMD method can and cannot do. Still, that knowledge is not yet systematic but rather fragmentary.

Findings on the protocol for LMD measurements and the analysis of the data cannot be transferred directly from one species to another or from one experimental set-up to another. Some research groups have developed their own working protocol, which they have applied in subsequent studies. However, several protocols exist side by side, making it difficult to compare the results of different research groups. Therefore, further research is needed, preferably in an integrated group of experts who develop a joint protocol for measurement and data analysis and to make it available to all others. This would also be very useful for researchers who want to use the LMD for the first time and who do not want to or cannot develop their own protocol. In this way, more CH_4 emissions from ruminants worldwide could be measured under on-farm conditions in the future, which would facilitate the development of mitigation measures for CH_4 and would enhance the efficiency of ruminant production.

Funding: This research received no external funding.

Institutional Review Board Statement: Not applicable.

Acknowledgments: I would like to thank Jelena Kecman and Alexander Finger for their valuable support in the preparation of the manuscript.

Conflicts of Interest: The author declares no conflict of interest.

Appendix A

Table A1. Overview of relevant studies with the laser methane detector.

Aim	No [1]	Dis [2]	Dur [3]	Rep [4]	Selected Result(s)	Ref [5]
Comparison of LMD [6] and RC [7]	20 dairy cows	1 m	5 min	3	Correlation between daily CH_4 g/kg energy corrected milk with LMD and RC: 0.82	[51]
Development of protocol for measurement and data analysis; nutritional study	2 + 12 dairy cows	0.6–1.2 m	2–3 min	26	Correlation of daily CH_4 production with RC and LMD: 0.55; natural grassland hay not recommended for indoor-fed cows	[18]
Relationship of blood serum parameters with CH_4 emission	46 ewes	1 m	3 min	-	Zinc, β-hydroxybutyrate, and non-esterified fatty acid blood concentrations had a partially breed-specific significant impact on CH_4 emission	[58]
Comparison of LMD, RC and MIR [8];	30 dairy cows	1 m	6 min	12	No agreement of LMD with RC or MIR	[17]
Book chapter reviewing previous work with the LMD	NA [9]	NA	NA	NA	Discussion of advantages and challenges of the LMD	[27]
Development of protocol for data analysis	2 dairy cows	3 m	5 min	15	Number of peaks best phenotype to discriminate high and low emitters	[42]
Comparison of CH_4 emissions from different husbandry systems	448 dairy cows	1 m	2 min	3	CH_4 concentrations were affected by location, breed, and husbandry system along a rural-urban gradient	[19]
Genetic analyses	330 ewes	1 m	3 min	-	h^2 [10] for CH_4 concentration 0.00–0.03	[15]
Development of protocol for measurement and data analysis	4 goats	2 m	4–5 min	60	No influence of restraint on CH_4 concentration; restraint of grazing goats can facilitate LMD measurements	[31]
Development of data analysis	46 dairy cows	1 m	5–6 min	6–8	Correlation of daily CH_4 production with RC, and several LMD phenotypes: up to 0.6; simple phenotypes not outperformed by those considering the time series nature of data	[54]
Comparison of LMD and other techniques for CH_4 measurement	NA	NA	NA	NA	Review and meta-analysis; sufficient correlation between methods for methods to be combined for international genetic studies	[10]
Comparison of LMD and NDIR [11] sniffer	48 dairy cows	1 m	5 min	6	Moderate agreement between LMD and NDIR sniffer	[46]
Effect of a novel feed additive on CH_4	30 dairy cows	1 m	4 min	-	Feed supplement significantly reduced CH_4 concentration ($p < 0.05$)	[16]
Nutritional study	45 dairy cows	1 m	-	-	Daily CH_4 production and CH_4 per kg milk decreased by grass-feeding compared to a total mixed ration	[57]
Breed comparison	24 beef cows	3 m	1 min	72	No between-breed difference in CH_4 output	[32]
Genetic analyses	622 dairy cows	2 m	5 min	1–3	h^2 for CH_4 phenotypes 0.05–0.28	[14]
Development of protocol for measurement and data analysis	4 + 12 goats	1 m	2 min	24 + 63	Recording interval should be 0.1 s; high variability of CH_4 concentration across individual goats and days; LMD able to detect diurnal pattern of CH_4 production in goats	[26]
Development of data analysis, comparison of LMD, GF [12] and NDIR/FTIR [13] sniffer	156 dairy cows	1, 2, 2.5 m	5 min	1–6	Number of CH_4 peaks = respiratory rate; similar ranking by LMD, GF and NDIR/FTIR sniffer sensors, regression equation for CH_4 g/d from LMD data	[35]
Nutritional study	18 goats	1.5 m	5 min	-	CH_4 production affected by energy and tannin levels of feed, and sex	[50]

Table A1. *Cont.*

Aim	No [1]	Dis [2]	Dur [3]	Rep [4]	Selected Result(s)	Ref [5]
Development of protocol for measurement and data analysis	71 + 18 dairy cows	1 m	~5 min	NA	Optimal recording duration > 3 min, correlation of LMD and RC CH_4 average 0.47, regression equation for CH_4 g/d from LMD data	[33]
Development of protocol for measurement and data analysis	8 dairy cows	3 m	1 min	40	The proposed LMD measurement protocol could not be successfully implemented due to local weather conditions (grazing)	[43]
Development of protocol for measurement and data analysis	-	0.4–2.5 m	5–10 min	NA	Duration of recording should be >2 min; no significant change in CH_4 concentration above 0.4 m distance; pointing angle, operator, LMD unit, time of day have significant effect on CH_4 concentration; number of peaks is the respiratory rate; prediction equation for total daily CH_4 from LMD record	[34]
Comparison of LMD and RC, comparison of 2 LMD devices	4 dairy cows	NA	continuous	NA	Good agreement in concentration in spent air from RC measured with RC and LMD, good agreement between 2 LMD devices	[40]
Genetic analyses	57 dairy cows	1 m	1–5 min	Up to 9	h^2 for CH_4 concentration 0.05	[13]
Development of protocol for measurement and data analysis	32 sheep	1 m	5–10 min	9–12	Correlation of daily CH_4 production with RC and LMD: 0.57	[29]
Comparison of Bonsmara, Nguni and Jersey cattle	12 heifers	3 m	1 min	40	Differences in CH_4 concentration between breeds and feed sources	[30]
Development of protocol for measurement and data analysis	24 ewes; 72 steers	?	2 min; 4 min	5; 3	Low correlation of CH_4 from RC and LMD-model improved by DMI [14]; best LMD phenotypes: length of eructation and maximum of respiration CH_4	[38]
Review summarizing previous work with the LMD	NA	NA	NA	NA	The LMD is a very recent tool and has potential to measure enteric CH_4 production in ruminants, further validation needed	[44]
Development of protocol for measurement and data analysis	2 + 24 dairy cows; 4 sheep	2.75 m	5 min	-	LMD has good sensitivity and specificity in detecting high and low CH_4 concentrations as compared to RC; cow activity and meteorological factors affect CH_4 concentrations	[36]
Development of protocol for measurement and data analysis	72 cross-bred steers	1 m	4 min	-	Significant regression of LMD on RC with R^2 = 0.27; similar results for mean of all values, all peaks, respiration peaks or eructation peaks	[28]
Comparison of LMD and RC	10 dairy cows	2.3 m	continuous (12–16 h)	NA	Good agreement in concentration in spent air from RC measured with RC and LMD	[45]
Development of protocol and data analysis	71 dairy cows	3 m	15–25 s	3	LMD is applicable for cows, data make biological sense	[12]

[1] Number and type of animals included in the study, [2] distance of measurement, [3] duration of recording, [4] repeats per animal, [5] references, [6] laser methane detector, [7] respiration chamber, [8] mid-infrared spectroscopy, [9] not applicable, [10] heritability, [11] non-dispersive infrared, [12] GreenFeed, [13] Fourier-transformed infrared, [14] dry matter intake.

References

1. Gerber, P.J.; Steinfeld, H.; Henderson, B.; Mottet, A.; Opio, C.; Dijkman, J.; Falcucci, A.; Tempio, G. *Tackling Climate Change through Livestock: A Global Assessment of Emissions and Mitigation Opportunities*; Food and Agriculture Organization of the United Nations (FAO): Rome, Italy, 2013; p. xxi + 115.
2. Knapp, J.R.; Laur, G.L.; Vadas, P.A.; Weiss, W.P.; Tricarico, J.M. Invited review: Enteric methane in dairy cattle production: Quantifying the opportunities and impact of reducing emissions. *J. Dairy Sci.* **2014**, *97*, 3231–3261. [CrossRef]
3. Kumari, S.; Fagodiya, R.; Hiloidhari, M.; Dahiya, R.; Kumar, A. Methane production and estimation from livestock husbandry: A mechanistic understanding and emerging mitigation options. *Sci. Total Environ.* **2020**, *709*, 136135. [CrossRef] [PubMed]
4. Reisinger, A.; Clark, H.; Cowie, A.L.; Emmet-Booth, J.; Gonzalez Fischer, C.; Herrero, M.; Howden, M.; Leahy, S. How necessary and feasible are reductions of methane emissions from livestock to support stringent temperature goals? *Philos. Trans. R. Soc. A* **2021**, *379*, 20200452. [CrossRef]
5. Zhao, Y.; Nan, X.; Yang, L.; Zheng, S.; Jiang, L.; Xiong, B. A review of enteric methane emission measurement techniques in ruminants. *Animals* **2020**, *10*, 1004. [CrossRef] [PubMed]

6. Zimmerman, P.R. Method and System for Monitoring and Reducing Ruminant Methane Production. U.S. Patent 7,966,971, 28 June 2011.
7. Lassen, J.; Løvendahl, P.; Madsen, J. Accuracy of noninvasive breath methane measurements using Fourier transform infrared methods on individual cows. *J. Dairy Sci.* **2012**, *95*, 890–898. [CrossRef] [PubMed]
8. Negussie, E.; Lehtinen, J.; Mäntysaari, P.; Bayat, A.R.; Liinamo, A.-E.; Mäntysaari, E.; Lidauer, M. Non-invasive individual methane measurement in dairy cows. *Animal* **2017**, *11*, 890–899. [CrossRef]
9. Johnson, K.; Huyler, M.; Westberg, H.; Lamb, B.; Zimmerman, P. Measurement of methane emissions from ruminant livestock using a sulfur hexafluoride tracer technique. *Environ. Sci. Technol.* **1994**, *28*, 359–362. [CrossRef] [PubMed]
10. Garnsworthy, P.C.; Difford, G.F.; Bell, M.J.; Bayat, A.R.; Huhtanen, P.; Kuhla, B.; Lassen, J.; Peiren, N.; Pszczola, M.; Sorg, D. Comparison of methods to measure methane for use in genetic evaluation of dairy cattle. *Animals* **2019**, *9*, 837. [CrossRef] [PubMed]
11. Wu, L. Measurement Methods to Assess Methane Production of Individual Dairy Cows in a Barn. Ph.D. Thesis, Wageningen University and Research, Wageningen, The Netherlands, 2016.
12. Chagunda, M.G.G.; Ross, D.; Roberts, D.J. On the use of a laser methane detector in dairy cows. *Comput. Electron. Agric.* **2009**, *68*, 157–160. [CrossRef]
13. Pickering, N.K.; Chagunda, M.G.G.; Banos, G.; Mrode, R.; McEwan, J.C.; Wall, E. Genetic parameters for predicted methane production and laser methane detector measurements. *J. Anim. Sci.* **2015**, *93*, 11–20. [CrossRef]
14. Mühlbach, S.; Sorg, D.; Rosner, F.; Kecman, J.; Swalve, H. Genetic analyses for CH_4 concentrations in the breath of dairy cows measured on-farm with the Laser Methane Detector. In Proceedings of the World Congress on Genetics Applied to Livestock Production, Auckland, New Zealand, 11–16 February 2018.
15. Reintke, J.; Brügemann, K.; Yin, T.; Engel, P.; Wagner, H.; Wehrend, A.; König, S. Assessment of methane emission traits in ewes using a laser methane detector: Genetic parameters and impact on lamb weaning performance. *Arch. Anim. Breed.* **2020**, *63*, 113–123. [CrossRef]
16. Vrancken, H.; Suenkel, M.; Hargreaves, P.R.; Chew, L.; Towers, E. Reduction of Enteric Methane Emission in a Commercial Dairy Farm by a Novel Feed Supplement. *Open J. Anim. Sci.* **2019**, *9*, 286–296. [CrossRef]
17. Denninger, T.M.; Schwarm, A.; Dohme-Meier, F.; Münger, A.; Bapst, B.; Wegmann, S.; Grandl, F.; Vanlierde, A.; Sorg, D.; Ortmann, S.; et al. Accuracy of methane emissions predicted from milk mid-infrared spectra and measured by laser methane detectors in Brown Swiss dairy cows. *J. Dairy Sci.* **2020**, *103*, 2024–2039. [CrossRef]
18. Kobayashi, N.; Hou, F.; Tsunekawa, A.; Yan, T.; Tegegne, F.; Tassew, A.; Mekuriaw, Y.; Mekuriaw, S.; Hunegnaw, B.; Mekonnen, W. Laser methane detector-based quantification of methane emissions from indoor-fed Fogera dairy cows. *Anim. Biosci.* **2021**, *34*, 1415–1424. [CrossRef] [PubMed]
19. Pinto, A.; Yin, T.; Reichenbach, M.; Bhatta, R.; Malik, P.K.; Schlecht, E.; König, S. Enteric Methane Emissions of Dairy Cattle Considering Breed Composition, Pasture Management, Housing Conditions and Feeding Characteristics along a Rural-Urban Gradient in a Rising Megacity. *Agriculture* **2020**, *10*, 628. [CrossRef]
20. Iseki, T. A portable remote methane detector using an InGaAsP DFB laser. *Environ. Geol.* **2004**, *46*, 1064–1069. [CrossRef]
21. Tokyo Gas Engineering Solutions. SA3C50A, LaserMethane Mini-g Operating Manual. Available online: https://www.crowcon.com/wp-content/uploads/2018/12/SA3C50A-Manual.pdf (accessed on 29 November 2021).
22. Crowcon Detection Instruments. LaserMethane Mini Specifications. Available online: https://www.crowcon.com/products/portables/lmm-gen-2/#specs (accessed on 26 November 2021).
23. Tokyo Gas Engineering Solutions. GasViewer, Laser Methane Linked Software for Android Devices, Instruction Manual. Available online: https://www.crowcon.com/wp-content/uploads/2018/12/GasViwer-Eng-manual.pdf (accessed on 26 November 2021).
24. Murray, R.; Bryant, A.; Leng, R. Rates of production of methane in the rumen and large intestine of sheep. *Br. J. Nutr.* **1976**, *36*, 1–14. [CrossRef] [PubMed]
25. Muñoz, C.; Yan, T.; Wills, D.A.; Murray, S.; Gordon, A.W. Comparison of the sulfur hexafluoride tracer and respiration chamber techniques for estimating methane emissions and correction for rectum methane output from dairy cows. *J. Dairy Sci.* **2012**, *95*, 3139–3148. [CrossRef]
26. Roessler, R.; Chefor, F.; Schlecht, E. Using a portable laser methane detector in goats to assess diurnal, diet- and position-dependent variations in enteric methane emissions. *Comput. Electron. Agric.* **2018**, *150*, 110–117. [CrossRef]
27. Jonker, A.; Chagunda, M.G.G.; Grobler, M.; Waghorn, G.C.; Van Wyngaard, J.D. Hand-held laser methane detector to determine methane plume concentrations from ruminants. In *Guideline for Estimating Methane Emissions from Individual Ruminants Using: GreenFeed, Sniffers, Hand-Held Laser Detector and Portable Accumulation Chambers*; Jonker, A., Waghorn, G.C., Eds.; MPI Technical Paper; Ministry for Primary Industries: Wellington, New Zealand, 2020; pp. 41–47.
28. Rooke, J.; Ricci, P.; Duthi, C.; Roehe, R.; Waterhouse, A. Measurements of methane using the laser methane detector are related to total daily methane output in beef cattle. In *Emissions of Gas and Dust from Livestock*; Hassouna, M., Guingand, N., Eds.; IFIP—Institut Technique du Porc: Paris, France, 2013; pp. 367–370. ISBN 978-2-85969-221-6.
29. Doran, S. Development of Proxy Indicators of Methane Output by Sheep Using Rapid-Throughput Field and Laboratory Technologies. Ph.D. Thesis, Aberystwyth University, Aberystwyth, UK, 2014.
30. Grobler, S.; Scholtz, M.; Van Rooyen, H.; Mpayipheli, M.; Neser, F. Methane production in different breeds, grazing different pastures or fed a total mixed ration, as measured by a Laser Methane Detector. *S. Afr. J. Anim. Sci.* **2014**, *44*, 12–16. [CrossRef]

31. Roessler, R.; Schlecht, E. Application of the laser methane detector for measurements in freely grazing goats: Impact on animals' behaviour and methane emissions. *Animal* **2021**, *15*, 100070. [CrossRef] [PubMed]
32. Mapfumo, L.; Grobler, S.; Mupangwa, J.; Scholtz, M.; Muchenje, V. Enteric methane output from selected herds of beef cattle raised under extensive arid rangelands. *Pastoralism* **2018**, *8*, 1–8. [CrossRef]
33. Bruder, T.; Rouillé, B.; Yan, T.; Chagunda, M.G.G. Towards a robust protocol for enteric methane measurements using a hand held Laser Methane Detector® in Ruminants. *ICAR Tech. Ser.* **2017**, *22*, 27–33.
34. Sorg, D.; Mühlbach, S.; Kecman, J.; Swalve, H. Protocol for measuring CH_4 concentrations with the laser methane detector in the breath of cows. In Proceedings of the 68th Annual Meeting of the European Federation of Animal Science, Tallinn, Estonia, 28 August–1 September 2017; p. 200.
35. Sorg, D.; Difford, G.F.; Mühlbach, S.; Kuhla, B.; Swalve, H.H.; Lassen, J.; Strabel, T.; Pszczola, M. Comparison of a laser methane detector with the GreenFeed and two breath analysers for on-farm measurements of methane emissions from dairy cows. *Comput. Electron. Agric.* **2018**, *153*, 285–294. [CrossRef]
36. Chagunda, M.G.G.; Ross, D.; Rooke, J.; Yan, T.; Douglas, J.-L.; Poret, L.; McEwan, N.R.; Teeranavattanakul, P.; Roberts, D.J. Measurement of enteric methane from ruminants using a hand-held laser methane detector. *Acta Agr. Scand. A-An.* **2013**, *63*, 68–75. [CrossRef]
37. Falconer, D.; Mackay, T.F. *Introduction to Quantitative Genetics*, 4th ed.; Longman Group Ltd.: Harlow, UK, 1996.
38. Ricci, P.; Chagunda, M.G.G.; Rooke, J.; Houdijk, J.G.M.; Duthie, C.-A.; Hyslop, J.; Roehe, R.; Waterhouse, A. Evaluation of the laser methane detector to estimate methane emissions from ewes and steers. *J. Anim. Sci.* **2014**, *92*, 5239–5250. [CrossRef]
39. Della, M.R.; Jonker, A.; Waghorn, G. A review of technical variations and protocols used to measure methane emissions from ruminants using respiration chambers, SF6 tracer technique and GreenFeed, to facilitate global integration of published data. *Anim. Feed Sci. Technol.* **2021**, *279*, 115018. [CrossRef]
40. Sorg, D.; Mühlbach, S.; Rosner, F.; Kuhla, B.; Derno, M.; Meese, S.; Schwarm, A.; Kreuzer, M.; Swalve, H. The agreement between two next-generation laser methane detectors and respiration chamber facilities in recording methane concentrations in the spent air produced by dairy cows. *Comput. Electron. Agric.* **2017**, *143*, 262–272. [CrossRef]
41. Bell, M.J.; Craigon, J.; Saunders, N.; Goodman, J.R.; Garnsworthy, P.C. Does the diurnal pattern of enteric methane emissions from dairy cows change over time? *Animal* **2018**, *12*, 2065–2070. [CrossRef]
42. Niero, G.; Cendron, F.; Penasa, M.; De Marchi, M.; Cozzi, G.; Cassandro, M. Repeatability and Reproducibility of Measures of Bovine Methane Emissions Recorded using a Laser Detector. *Animals* **2020**, *10*, 606. [CrossRef]
43. Van Wyngaard, J.; Du Toit, C.; Meeske, R.; Erasmus, L.; Goemans, Z. Laser methane detector: Not a practical measurement tool for grazing dairy systems. In Proceedings of the South African Society for Animal Science, Port Elizabeth, South Africa, 18–21 September 2017; pp. 244–245.
44. Chagunda, M.G.G. Opportunities and challenges in the use of the Laser Methane Detector to monitor enteric methane emissions from ruminants. *Animal* **2013**, *7* (Suppl. S2), 394–400. [CrossRef]
45. Chagunda, M.G.G.; Yan, T. Do methane measurements from a laser detector and an indirect open-circuit respiration calorimetric chamber agree sufficiently closely? *Anim. Feed Sci. Technol.* **2011**, *165*, 8–14. [CrossRef]
46. Rey, J.; Atxaerandio, R.; Ruiz, R.; Ugarte, E.; González-Recio, O.; Garcia-Rodriguez, A.; Goiri, I. Comparison between non-invasive methane measurement techniques in cattle. *Animals* **2019**, *9*, 563. [CrossRef] [PubMed]
47. Tukey, J.W. *Exploratory Data Analysis*; Addison-Wesley Publishing Company: Boston, MA, USA, 1977.
48. Meinert, T.R.; Norman, H.D. Merit of Outliers for Milk Yield as Indicators of Accuracy of Genetic Evaluations of Sires. *J. Dairy Sci.* **1998**, *81*, 2951–2955. [CrossRef]
49. Schmidt-Nielsen, K. *Animal Physiology: Adaptation and Environment*, 5th ed.; Cambridge University Press: Cambridge, UK, 1997.
50. Washaya, S.; Mupangwa, J.; Muchenje, V. Chemical composition of Lablab purpureus and Vigna unguiculata and their subsequent effects on methane production in Xhosa lop-eared goats. *S. Afr. J. Anim. Sci.* **2018**, *48*, 445–458. [CrossRef]
51. Kecman, J.; Heinicke, J.; Grütter, L.; Kuhla, B.; Swalve, H.H. Methane emissions from dairy cows measured using laser methane detector and respiration chambers. In Proceedings of the 72nd Annual Meeting of the European Federation of Animal Science, Davos, Switzerland, 30 August–3 September 2021; p. 123.
52. Derno, M.; Elsner, H.G.; Paetow, E.A.; Scholze, H.; Schweigel, M. Technical note: A new facility for continuous respiration measurements in lactating cows. *J. Dairy Sci.* **2009**, *92*, 2804–2808. [CrossRef] [PubMed]
53. Llonch, P.; Troy, S.M.; Duthie, C.-A.; Somarriba, M.; Rooke, J.; Haskell, M.J.; Roehe, R.; Turner, S.P. Changes in feed intake during isolation stress in respiration chambers may impact methane emissions assessment. *Anim. Prod. Sci.* **2016**, *58*, 1011–1016. [CrossRef]
54. Brocklehurst, S.; Hargreaves, P.; March, M. Novel statistical methods to improve analysis of laser methane detector data. *Proc. Br. Soc. Anim. Sci. Adv. Anim. Biosci.* **2019**, *10*, 1–284. [CrossRef]
55. de Haas, Y.; Veerkamp, R.F.; de Jong, G.; Aldridge, M.N. Selective breeding as a mitigation tool for methane emissions from dairy cattle. *Animal* **2021**, *15*, 100294. [CrossRef]
56. Lassen, J.; Difford, G.F. Genetic and genomic selection as a methane mitigation strategy in dairy cattle. *Animal* **2020**, *14*, s473–s483. [CrossRef] [PubMed]

57. Cameron, L.; Chagunda, M.; Roberts, D.; Lee, M. A comparison of milk yields and methane production from three contrasting high-yielding dairy cattle feeding regimes: Cut-and-carry, partial grazing and total mixed ration. *Grass Forage Sci.* **2018**, *73*, 789–797. [CrossRef]
58. Reintke, J.; Brügemann, K.; Yin, T.; Wagner, H.; Wehrend, A.; Müller, A.; König, S. Associations between minerals and metabolic indicators in maternal blood pre-and postpartum with ewe body condition, methane emissions, and lamb body weight development. *Animal* **2021**, *15*, 100034. [CrossRef] [PubMed]
59. Wu, L.; Koerkamp, P.W.G.; Ogink, N. Uncertainty assessment of the breath methane concentration method to determine methane production of dairy cows. *J. Dairy Sci.* **2018**, *101*, 1554–1564. [CrossRef] [PubMed]

MDPI AG
Grosspeteranlage 5
4052 Basel
Switzerland
Tel.: +41 61 683 77 34

Methane Editorial Office
E-mail: methane@mdpi.com
www.mdpi.com/journal/methane

www.ingramcontent.com/pod-product-compliance
Lightning Source LLC
Chambersburg PA
CBHW062037220926
51850CB00010B/2585

* 9 7 8 3 7 2 5 8 6 8 7 6 6 *